Economic Research Relevant to the Formulation of National Urban Development Strategies

Part One

James Douglas McCallum, Lowdon Wingo, Wilbur Thompson. H.W. Richardson, Joel Bergsman, Peter Greenston and Robert Healy

First published in 1971
by Resources for the Future Press

This edition first published in 2017 by Routledge
2 Park Square, Milton Park, Abingdon, Oxon, OX14 4RN
and by Routledge
711 Third Avenue, New York, NY 10017

Routledge is an imprint of the Taylor & Francis Group, an informa business

ISBN 13: 978-1-138-18485-5 (hbk)
ISBN 13: 978-1-315-64490-5 (ebk)
ISBN 13: 978-1-138-18486-2 (pbk)

Routledge Revivals

Economic Research Relevant to the Formulation of National Urban Development Strategies

Originally published in 1971, this volume contains papers invited for a conference on economic research relevant to national urban development held in September of the same year. The conference pulled together researchers from both the United Kingdom and the United States who were interested in economic research on key issues of both countries' management of their urban areas. Papers are varied from those in the early stages of research to those whose research has been completed and all provide an insight into the increase of urbanisation present in the first world. This title will be of interest to students of environmental studies and economics.

PREFACE

This volume is one of two consisting of 16 papers invited for the Conference on Economic Research Relevant to National Urban Development Strategies jointly sponsored by Resources for the Future, Inc. and the University of Glasgow, at Wolfson Hall of the University August 30 - September 3, 1971.

The Conference was designed to bring together a small group of economists and related professionals from the United Kingdom and the United States who are engaged in economic research having a strategic bearing on key issues being confronted by the two countries in managing their urban development. It is hypothesized that a generation of British policy experience and analysis may illuminate some of the problems of formulating urban development policies in the U.S., while the rapidly growing volume of urban economic research and analysis in the U.S. may generate some insights into problems of the urban economy in the United Kingdom. Finally, it is hoped that such an intensive discussion can sharpen priorities for a succeeding generation of urban economic research on both sides of the Atlantic.

Because of the nature of the Conference the papers vary considerably in style and content; some are early research reports, other reconnoitre the policy field, some are relatively conclusive reports on recently completed research efforts. In all, they provide a view of the way the state of knowledge about urbanization in advanced countries is expanding at the frontier of urban economics.

Lowdon Wingo
Resources for the Future

Gordon Cameron
University of Glasgow

June 24, 1971

Conference Co-Chairman

CONTENTS

21 May 1971

A REVIEW HISTORY OF BRITISH REGIONAL POLICY

James Douglas McCallum

Department of Town & Regional Planning

University of Glasgow

Glasgow, W.2, Scotland

NOTE: This paper has been prepared for the Resources for the Future - University of Glasgow Conference on Economic Research Relevant to National Urban Development Strategies, Glasgow, Scotland, August 30 - September 3, 1971. It is subject to revision and should not be cited or quoted without express permission of the author.

JAMES DOUGLAS McCALLUM

Biographical Statement

Born in Memphis, Tennessee, Mr. McCallum was educated at the Massachusetts Institute of Technology (B.Sc., 1964, Politics and Economics) and at the University of North Carolina at Chapel Hill (M.R.P., 1967, and Ph.D., pending, both in Regional Planning). He has worked in planning in the U.S.A., and in 1966 was an O.A.S. Research Fellow in Bogotá, Colombia. He continued his researches in 1968-1969 in London, at the Joint Unit for Planning Research. Since September 1969 he has been teaching in the Department of Town and Regional Planning at the University of Glasgow.

A REVIEW HISTORY OF BRITISH REGIONAL POLICY

British regional policy has a long history, having developed almost continuously since 1934 despite changes in the nation's economic situation and shifts in the attitude of incumbent Governments.[1] An impressive stream of Commission Reports, White Papers, and Acts of Parliament -- as well as books, monographs, theses, and articles -- marks the progress of this development. The past eight or nine years, however, have seen a considerable acceleration of this process, and Britain now has probably the most active, extensive, and expensive body of regional development policy in the West.[2] But because of the continuity of this gradual policy build-up -- and also because of the persistence of regional problems -- it is important to examine British policy in its historical perspective.

The Beginnings of Regional Policy

The existence of a 'regional problem' became generally recognized during the Depression years, as large regional differences in unemployment rates persisted throughout the late Twenties and early Thirties. The first government response, the Industrial Transference Board and its related system of training centers, was set up in 1928 to assist voluntary labor migration out of the hard-hit regions. These efforts had little effect, however, because of the over-all surplus: aggregate national demand simply could not absorb the labor supply, regardless of geographical rearrangements (McCrone, 1969, pp 91-102; Odber, 1965, pp 332-334). The numbers involved in transference schemes were relatively small, and migration sometimes proved temporary, as many as one-quarter of transferees eventually returning to their original homes.[3] The Liberal Party in 1929 tried to generate an interest in doing something about the regional situation -- its proposals largely the work of Keynes -- but failed to secure either the support of the electorate or the concern of the ruling Government.

Conditions continued to worsen, and by 1934 unemployment rates in the North of England and in Scotland averaged about 22 per cent and in Wales around 30 per cent -- in contrast to some 10 or 11 per cent in the South and Midlands; some small areas within the depressed regions had local jobless rates of 40, 60 or even 80 per cent! The findings of a 1934 official investigation of the situation substantiated the extremity of conditions in many areas, particularly those heavily dependent upon ship-building, coal-mining, metal manufacture, heavy engineering, and textiles.[4]

The investigators judged that although there was considerable surplus labor, the potential of migration was probably minor. More important, they concluded that depressed areas

> "can only escape from the vicious cycle, where depression had created unemployment and unemployment intensified depression, by means of some positive external assistance."
> (Ministry of Labour, 1934, p 106)

The Government then brought forth a measure of "positive external assistance" with the Special Areas (Development and Improvement) Act of 1934, Britain's first explicitly regional legislation.[5] Four depressed areas were officially delimited, and two Commissioners were set up to administer the new legislation; it was an "extremely cautious" beginning (Loasby, 1965, pp 34-35). Initially the Commissioners had little power or money, and in the first year "expenditure was almost entirely limited to sewerage schemes and settlement of labour on the land."[6] (McCrone, 1969, p 95) But in 1936 loan capital was made available for small businesses in the Special Areas (as the designated areas were then called), this fund being extended in 1937. The 1937 Special Areas (Amendments) Act introduced a tax incentive, and the Commissioners were also empowered to make contributions toward off-setting a firm's taxes, rent, or rates (local property taxes).[7]

The first industrial estates had been established in 1936, and this development was officially encouraged by incorporation into the 1937 Amendments. Several of the early estates then set up became moderate successes even before the War, though some writers (Odber, 1957, 1965; Labour Party, 1970) attribute this to the enterprise and activity of European emigrants who fled to Britain during the late Thirties.[8] However, the industrial estate concept itself has proved immensely durable, featuring prominently in regional development policies almost continuously from 1937 to the present.

By the 1939 the Special Areas Commissioners had built up a fairly broad range of powers and gained considerable experience in dealing with the problems of depressed regions (Ministry of Labour, 1935, 1936a, 1936b, 1937, 1938; Scottish Office, 1935, 1936a, 1936b, 1937, 1938). And though their activities had produced little visible economic effect (unemployment only began to drop substantially in 1939 with the impact of the war economy), it must be remembered that most of the important powers came into effect only in 1937 or 1938 and so did not have time to prove themselves (Rodwin, 1970, pp 112-113). Still, the proportion of new factories going to the Special Areas -- about 5 per cent up through 1937 -- jumped to 17 per cent in 1938, clearly an indication of some success.

Moreover, critical discussion from the Commissioners themselves as well as from outside observers such as Dennison (1939) began, in the late Thirties, to provide some evaluation of regional development policies. It was pointed out, for example, that the delineation of the Special Areas was less than fully satisfactory; they tended to be only parts of larger regions, and in some cases the worst-hit areas were not included. In addition, the designated areas generally excluded major urban centers (such as Glasgow, Cardiff, Swansea, and Newcastle) and thus excluded the "natural focal points" of development (McCrone, 1969, p 93).

Partly because of such uncertainties about existing regional policy -- but primarily because of the continuing concern, only heightened by the distress of the Depression, about the quality of urban life in the large industrial centers of the nation -- a Royal Commission was established in July 1937

> "to inquire into the causes which have influenced the present geographical distribution of the industrial population of Great Britain and the probable direction of any change in that distribution in the future; to consider what social, economic or strategical disadvantages arise from the concentration of industries or of the industrial population in large towns or in particular areas of the country; and to report what remedial measures if any should be taken in the national interest." (Royal Commission on the Distribution of the Industrial Population, 1940, p 1)

The Commission's Report, now well known as the "Barlow Report", was completed in August 1939 and published the following January.

Many of the key issues and concepts of the Barlow Report had appeared previously in policy discussions. The 1934 investigators, for example, had already sounded two themes which were to appear in Barlow: first, a firm conviction that the net effects of large conurbations were "bad"[9]; and second, the explicit acceptance of positive governmental control and direction of private industry as a necessary means of securing a desired regional pattern.

> "The evils, actual and potential, of this increasing agglomeration of human beings are so generally recognized as to need no comment. It is suggested, therefore, that the time has come when the Government can no longer regard with indifference a line of development which, while it may possess the initial advantage of providing more employment, appears on a long view to be detrimental to the best interests of the country; and the first practical step which could be taken toward exercising a measure of control in this direction would seem to be some form of national planning of industry."
> (Ministry of Labour, 1934, p 107)

Sir Malcolm Stewart (Special Areas Commissioner for England and Wales) had also put forward some of these same views, including a strong argument that Depressed Area problems could not be solved in isolation or without coordinated control over development in the more prosperous regions (Ministry of Labour, 1936b). Nonetheless, the Barlow Report covered considerable new ground, discussing the issues so thoroughly that McCrone (writing 29 years later) could call it "...the most comprehensive review yet to be undertaken in Britain of the case for a regional policy." (McCrone, 1969, p 104)

In its Report the Barlow Commission listed three principal "objectives of national action":

> "(a) Continued and further redevelopment of congested urban areas, where necessary.
> (b) Decentralization or dispersal, both of industries and industrial population, from such areas.
> (c) Encouragement of a reasonable balance of industrial development, so far as possible, throughout the various divisions or regions of Great Britain, coupled with appropriate diversification of industry in each division or region throughout the country." (Royal Commission on the Distribution of the Industrial Population, 1940, pp 201-202)

To implement their proposals for accomplishing these objectives the Commission recommended the establishment of a national authority with broad new powers. They split, however, over the nature of this new body, a minority calling for somewhat more extensive and active powers than the majority. The case for a system of national development controls to restrict growth in the South East was also strongly urged, with a minority statement calling for such powers to be applied over the whole country.

Regardless of the fate of the Report's specific recommendations, its several hundred pages of discussion directly addressed to questions of urbanization and regional development vastly broadened the public debate on these problems and provided a solid basis for future policy deliberations. It is in this respect that McCrone gives the Report such an important place in the history of British regional policy:

> "...the Barlow Report was a landmark in the development of thought on the regional problem in Britain. ... In several respects it was ahead of its time and many of the innovations which have been introduced into British regional policy in the 1960s were given clear expression in this report some twenty years earlier." (McCrone, 1969, p 104)

Still, the Report had its short-comings. For one thing, the Commission interpreted its terms of reference so as to exclude most of the relevant discussion of regional economic "imbalances"; most of the members apparently felt that remedial measures such as those in existence by 1938/39 would eventually suffice to return matters to a more "balanced" regional situation. Also, there was seemingly no clear recognition in the Report of the linkages between the two phenomena of congestion and decline, and so they were discussed independently, continuing that separation of urban (planning) concerns and regional (economic) concerns which characterized British thought until the 1960s. The main theme of the Commission was urban congestion and measures to counter it -- and not (directly, at least) regional economic development. For instance, the Report proposed garden cities, satellite towns, industrial estates, and development of existing small towns as measures to assist decentralization; any potential relevance of such devices for regional economic rather than local decongestion purposes was largely ignored -- a dichotomy in outlook which was later to have unfortunate consequences.[10]

Later during the war another major Committee reported, this time on land utilization in the rural areas. The "Scott Report", while somewhat limited in scope, was an important document and had, later, considerable impact on rural planning thought and policy (Ministry of Works and Planning, 1942).

The war had meantime temporarily postponed detailed consideration of the Barlow Report, though its influence was felt in the 1943 creation of the Ministry of Town and Country Planning. Following the Report's main emphasis (while stopping short of its recommended new central authority), the Ministry was largely concerned with urban problems and physical land use activities. A year later the 1944 White Paper on Employment Policy (Ministry of Reconstruction, 1944) picked up another theme from Barlow and devoted one of its six chapters to "The Balanced Distribution of Industry and Labour". In it the Government accepted both the objective of balanced distribution and the need to use a variety of powers and controls to secure such a development.[11] In only ten years Britain had thus come to the position of endorsing regional policy as a proper and major sphere of central government concern and activity.

The Post-War Period: To 1951

Two months after V-E Day a sweeping victory at the polls brought to power a Labour Government strongly committed to social change generally and to governmental planning specifically. Despite the economic pressures and problems of the time, the Government soon brought forth a substantial body of legislation on regional policy. The central piece was the

Distribution of Industry Act of 1945, successor to the Special Areas Acts. The designated geographical areas (now called Development Areas) were roughly the same as the former Special Areas -- with the significant addition that those major cities formerly excluded were now included.[12] The powers to influence industrial development, mostly given to the Board of Trade, were generally the same as in the Special Areas legislation, though without tax concessions.

Control over new development as proposed in Barlow was not included in this Act; prohibitory powers were proposed but were then dropped during the Bill's passage, and industrialists were only required to notify the Board of Trade of intended new developments over 10,000 square feet. Control was established, however, by deliberate discriminatory use of the powerful war-time building license system, which remained in force until 1954 and proved a stronger tool than the more modest powers originally proposed for the Act.

In March 1946 the Minister of Town and Country Planning announced that the Government's policy on Greater London, based on the Barlow proposals, would combine restraints on over-all growth with planned decentralization. In 1947 came the Town and Country Planning Acts, which explicitly recognized "the relationship between distribution of industry policy and town and country planning" (Board of Trade, 1948, p 12). This legislation also introduced the Industrial Development Certificate (I.D.C.), the permanent form of development control intended to replace the building license system. Industrial buildings in excess of 5,000 square feet were required to receive a certificate stating that the "development in question can be carried out consistently with the proper distribution of industry". The link with town and country planning was to be achieved by making I.D.C. approval (obtained from the Board of

Trade) mandatory before planning permission could be given by the local planning authority; and under land-use planning law this permission is necessary for almost all non-agricultural construction and development.

This general set of powers remained in force during the early post-war period, with the "stick" (building license and I.D.C. control, discriminatory provision of rationed scarce materials) and the "carrot" (financial incentives, ready-to-use factories) both being utilized. But another significant post-war innovation, the official New Towns policy, turned out in practice to have little direct impact on regional policy. The idea was generally interpreted (following Barlow) as relating only to intra-regional movements resulting from urban redevelopment.[13]

Initially, the economic environment was quite favorable for progress in the Development Areas. Conditions were buoyant nationally, the steady high demand for capital investment (and therefore industrial development) reflecting not only replacement of war-time damage and implementation of war-postponed plans, but also the impact of releasing the economy's considerable stored-up liquidity. The Development Areas at this time also contained a larger surplus of labor than other areas, because of both higher unemployment and lower participation rates. In addition, many of the Ordnance and other war-time factories which were being converted to peace-time industrial use had been located in the Development Areas through a policy of strategic dispersal, and the availability of such ready-to-use factories was an important attraction at that time. Finally, since many capital goods and building materials were in short supply and rationing still in force, the Government was able to use discriminatory allocation to help secure the desired movement of industry.

Helped by all these positive forces, the Development Areas initially gained the lion's share of new industrial investment. Measured by area of approved new buildings (over 5,000 square feet only), the Development Areas -- with under 20 per cent of the nation's population -- received over 50 per cent of the total national industrial growth in 1945-46-47, an average of about 15,700,000 square feet per year. This initial success was extremely significant, not so much because of the immediate benefits to the Development Areas, but because it fixed in many minds the firm conviction that regional policy, if stoutly applied, can work. Yet, perversely, this very success had the unfortunate short-term effect of making many people feel that the regional problems were "practically solved", and thus it eroded the feeling of urgency and paved the way psychologically for later reductions in Development Area priority.

In October 1947 the situation changed dramatically, much to the detriment of regional policy (Loasby, 1965, pp 37-39). In response to a balance of payments crisis the Government forced heavy cuts in construction and especially industrial building. These cuts were coupled with a shift of emphasis, assigning national priority to export (or import-substitution) industries. The general buoyancy of demand was equally a casualty of the crisis. Regional policy was quietly moved to a back burner where it was to remain, simmering, until the Sixties.

In line with these over-all cuts, the building of advance factories in the Development Areas was stopped -- another tool of regional policy which was not to be resurrected for many years. With the demands of exports so pressing, moreover, the Government could no longer stringently enforce a regionally discriminatory pattern with either building license or I.D.C. controls. Similarly, the power of control through rationed materials was lessened. By this time also the supply of war-time factories available for conversion had largely dried up, eliminating that tool of policy. Finally, the discrepancies in labor availability had been substantially reduced, weakening yet another of the Development Areas' attractions to industry.

In consequence, the Development Areas' share of approved industrial expansion in the period 1948-51 dropped sharply, down to 19.4 per cent of the total. Even in absolute terms the reverse was clear: the annual average area of approved new industrial development dropped from 15,734,000 square feet in 1945-47 to 8,602,000 in 1948-51 -- a drop of 45 per cent!

TABLE 1

NEW INDUSTRIAL BUILDING APPROVED[14]
(ANNUAL AVERAGE DURING PERIOD)

	Development Areas:		Rest of Great Britain:	
	Square Feet	% Of Total	Square Feet	% Of Total
1945-47	15,734,000	51.3 %	14,949,000	48.7 %
1948-51	8,602,000	19.4 %	35,819,000	80.6 %
1952-55	9,906,000	17.4 %	47,180,000	82.6 %

The Government did not, of course, simply ignore these trends in regional growth. But they felt that their hands were tied by the severity of external economic forces, particularly the balance of payments question; regional policy simply could not be given its former priority. In 1950, when the balance of payments improved following the devaluation of sterling, the Government did try to breathe new life into regional policy, and "the second Distribution of Industry Act (1950) was followed by a decision to relax slightly for certain locations of relatively high unemployment the stringent tests by which factory building was regulated." (Odber, 1957, p 19) A slight improvement in the regional situation began to emerge, but this was soon squashed when the "improved" national economic situation proved to have been only temporary.

Even considering these economic difficulties, though, it is clear that the Government no longer accorded regional policy its former place of importance. Expenditure on regional policies under the provisions of the 1945 Act, for example, dropped from some $68 million in 1947/48-1948/49 to about $35 million in 1949/50-1950/51.[15] In addition, the number of firms ready and able to expand or move had declined sharply as the war-time back-log was extinguished; the slow growth of the "normal" peace-time economy simply did not generate enough potential moves to maintain the earlier pace. Perhaps never again would the combination of economic situation and Governmental attitude be as favorable for the Development Areas as during 1945-1947.

The Post-War Period: 1951 to 1963

The 1951 General Election, bringing to power a Conservative Party which had campaigned on a "freedom from planning" platform, signalled the end of the first phase of regional policy activism. And while it might be said that Labour in the years 1948-51 had been forced somewhat against their inclination to downgrade regional policy, the Conservatives of 1951 could scarcely restrain their enthusiasm for dismantling the apparatus of governmental planning, including regional policy. While their ideological fervor was later moderated by the responsibilities and constraints of power, they still managed to cripple a number of promising and worth-while beginning efforts -- a destruction they would themselves be forced to remedy some ten or twelve years later.[16] For example, the regional offices which had been established in several Ministries to encourage and coordinate regionally-significant information and decisions were disbanded, and the few official regional research activities then in existence were discontinued (Harris, 1966b). The new Government also ceased designating new towns; and they took no action to revise existing regional plans, much less to prepare new ones for areas not previously covered.[17]

There was not, however, a great deal of change in legislation; the 1945, 1947, and 1950 Acts remained in force. But the locational control theoretically to be exercised through the I.D.C. system was visibly loosened; certificates became easier to obtain for non-Development Area locations and so the "stick" of regional policy was softened. Simultaneously, the "carrot" was allowed to wither. Expenditures on regional policy in the five year period 1946/47 through 1950/51 had averaged about $24.1 million per year, but in the eight year period 1951/52 through 1958/59 the expenditures averaged only about $11.4 million annually, despite a considerable expansion in the total national budget at the same time.[18]

The new Government was able to 'get away with' this relaxation mainly because of an intermittent boom in economic conditions which temporarily sustained demand in the traditional industries on which the Development Areas were so heavily dependent.[19] But no one seemed to have considered that these often troubled industries might suffer a relapse and when they did falter in the late Fifties the Government was largely unprepared. Having failed to secure any significant broadening of their industrial bases, the Development Areas suffered heavily in the ensuing decline. (See Table 2)

Unfortunately for the distressed areas, Governments (of both parties) had long tended to be insensitive to the regional consequences of national policies, partly because of their persistent mental compartmentalization of regional affairs and national affairs into separate and unconnected sets of phenomena -- a tendency aggravated by a similar attitude among most economists. National economic growth was generally slow throughout

TABLE 2

UNEMPLOYMENT RATES IN THE REGIONS
(AVERAGES OF MONTHLY FIGURES, PER CENTS)[20]

Regions	1949	1950	1951	1952	1953	1954	1955	1956	1957	1958	1959	1960	1961	1962
London & South E.	1.1	1.1	0.9	1.3	1.2	1.0	0.7	0.8	1.0	1.3	1.3	1.0	1.0	1.3
Eastern	1.1	1.2	0.9	1.3	1.3	1.2	0.9	1.0	1.3	1.6	1.5	1.2	1.1	1.4
Southern	1.4	1.4	1.1	1.4	1.4	1.1	0.9	1.0	1.3					
South Western	1.4	1.4	1.2	1.5	1.6	1.5	1.2	1.2	1.8	2.2	2.2	1.7	1.4	1.8
Midland	0.6	0.5	0.4	0.9	1.1	0.6	0.5	1.1	1.3	1.6	1.5	1.0	1.4	1.8
North Midland	0.6	0.6	0.5	1.0	0.7	0.6	0.5	0.6	1.0	1.5	1.5	1.1	1.0	1.4
E. & W. Yorkshire	0.9	0.9	0.9	1.9	1.2	0.9	0.7	0.8	0.9	1.9	1.9	1.2	1.0	1.6
North West	1.7	1.6	1.2	3.6	2.1	1.5	1.4	1.3	1.6	2.7	2.8	1.9	1.6	2.6
Northern	2.6	2.8	2.2	2.6	2.4	2.2	1.8	1.6	1.7	2.4	3.3	2.8	2.5	3.8
Scotland	3.0	3.0	2.5	3.3	3.1	2.8	2.4	2.4	2.6	3.7	4.4	3.6	3.1	3.8
Wales	4.0	3.7	2.7	2.9	3.0	2.5	1.8	2.0	2.6	3.8	3.8	2.7	2.6	3.8
GREAT BRITAIN	1.5	1.5	1.2	2.0	1.6	1.3	1.1	1.2	1.4	2.1	2.2	1.6	1.6	2.1

the mid- and late- Fifties, and Government economic policy was the now well-known "Stop-Go". In reaction to a balance of payments difficulty, the Government would impose restraint and deflation, which then brought unemployment, recession, and eventually a balance of payments improvement; this would be followed by fast reflation and rising inflation, usually bringing another balance of payments crisis and another round of deflation.[21] (Hutchison, 1968; Mitchell, 1966; Hagen & White, 1966) It is particularly during periods of recession or stagnation when regional problems become most severe and there is greatest need for special assistance to hard-hit areas; but from late 1947 when regional policy was first sacrificed to national economic needs, the Government's inevitable response continued to be just the reverse. The Treasury, particularly, as a stronghold of traditional conservative orthodoxy, always insisted on uniform tight squeezes and always viewed regional policy as an unwelcome luxury to be cut at the very first opportunity.[22]

The non-policy of 1951 through the mid-Fifties was partially compensated for by external economic forces and conditions which at least did not bring any visible decline in the poorer regions (though certainly not bringing any improvement in the discrepancies which already existed). But continuing this non-policy during the stagnant years, especially with the recession in 1958 and 1959, inevitably contributed to a sharp decline in the depressed regions.[23] Eventually forced by rising unemployment to reconsider its policies, the Government (headed since 1957 by Mr. Macmillan and distinctly less laissez-faire than its Conservative predecessors) responded with the Distribution of Industry (Industrial Finance) Act of 1958. A number of regions were newly added to the list of eligible areas, and the loans and grants to industry were made somewhat more widely available. In addition, the amount of money committed to regional purposes increased from $8.6 million in 1958/59 to $20.6 in 1959/60. But on the whole the powers and tools remained substantially as before. More crucially -- and despite the fact that momentarily more attention was being given to the regional problem -- these efforts were simply immediate responses to immediate problems, hastily conceived, poorly coordinated, and not consistently related to any coherent or clear national policy program or strategy for regional development.[24]

The regional problems persisted, however, and in two years the several previous pieces of legislation were all repealed and replaced by the Local Employment Act of 1960. Under this Act the Board of Trade was empowered to schedule 'Development Districts', the criterion for their selection being current unemployment rates. Most of the previously-designated areas were again included, with a number of rural areas (which had been largely ignored in earlier legislation) being added. But since the Board of Trade was given specific discretionary power to add or remove areas as they rose above or fell below the national unemployment criterion, the extent of the scheduled areas could -- and did -- change repeatedly and rapidly. In terms of percentage of national population, Development Districts represented 12.5 per cent in 1961, 7.2 per cent in 1962, and 16.8 per cent in 1966. This feature undermined the effectiveness of the program, since rapid and frequent changes in area coverage made public or private planning very difficult, the investor or developer never sure what financial assistance would be available for which areas at which times.

A few other modest changes in attitude emerged in the early Sixties. In 1962, for example, a number of separate departments were grouped together to form the Scottish Development Department, a move suggested by the Toothill Report (Scottish Council, 1961) to facilitate the "promotion and coordination of economic and industrial policy in Scotland".[25] This was a significant move, for it brought together for the first time into one powerful body both the infrastructure (physical) and development (economic) planning activities. October 1963 brought the first appointment of a Government Minister to be specifically concerned with regional development: the designation of the Secretary of State for Industry, Trade, and Regional Development.[26]

Thought was again being given to the regional problem, and tools and techniques were being examined critically, with an eye to devising new measures. This was partly motivated by the worsening regional situation and the related political ramifications of that deterioration for the incumbent Government in the approaching General Election.[27] It was also partially motivated by the new interest in the whole problem of national economic performance which finally brought a Conservative Government to embrace, however tentatively, the idea of national planning which its predecessors had so firmly and religiously rejected.[28] These meagre beginnings of a new period of activism and constructive thought began to bear fruit in 1963 with a number of documents which can be taken as the starting point of a "new era" in regional policy.

The "New Era" in Regional Policy: 1963 to 1970

Despite a wide-spread tendency to see regional development, physical planning, and the national economy as separate and distinct sets of issues and problems, some of the 1963 documents hinted at a new, and more integrated, approach. The National Economic Development Council (N.E.D.C.), set up to study and advise on the national economy, also addressed itself to regional problems. In addition, two 1963 White Papers, each a response to specific local distress in the region studied, revealed a broader view of regional development than was usual, considering both physical planning and national economic aspects.

Slightly under a year after it first met the N.E.D.C. produced a study of the "implications of an annual growth rate of 4 per cent for the period 1961-1966" (National Economic Development Council, 1963b). This was followed by a quasi-policy document Conditions Favourable to Faster

Growth (N.E.D.C., 1963a) which suggested some of the specific ways in which public and private sectors would need to perform if this faster national economic growth were to be achieved.

In addition to its principal themes, Conditions Favourable to Faster Growth contained not only a lengthy discussion of "Regional Questions" but also an explicit statement that regional policy should be considered something more than just special welfare measures:

> "The level of unemployment in different regions of the country varies widely, and high unemployment associated with the lack of employment opportunities in the less prosperous regions is usually thought of as a social problem. Policies aim, therefore, to prevent unemployment rising to politically intolerable levels and expenditure to this end is often considered as a necessary burden to the nation, unrelated to any economic gain that might accrue from it. But the relatively high unemployment rates in these regions also indicate considerable labour reserves. To draw these reserves into employment would make a substantial contribution to national employment and national growth." (N.E.D.C., 1963a, p 14)

Given this new view of unemployment as a "resource", an important interdependence between the national and the regional economies was then argued:

> "A national policy of expansion would improve the regional picture; and, in turn, a successful regional development programme would make it easier to achieve a national growth programme." (N.E.D.C., 1963a, p 29)

One of the regional policy issues discussed was the increasing congestion cost of development in the South East and the Midlands and the need for effective counter-measures. In this respect a suggestion was made for extending development controls to office buildings, a step taken shortly afterwards. There was also argument for a "growth area" approach, as well as a suggestion that larger policy regions would be more appropriate than those currently in use under the 1960 Act. Additionally, it was recommended that the level of government expenditure on physical infrastructure in the development areas be continued at a high level.

The White Papers for Central Scotland (Scottish Development Department, 1963) and for the North East (Board of Trade, 1963) appeared in late 1963 and represented, unlike the N.E.D.C. reports, official Government policy statements. Each investigation was the result of particular regional pressures upon the Government from economically hard-hit regions; they did not represent parts of any comprehensive regional study of the nation. Nonetheless, they did constitute a beginning step toward national regional planning, and they also introduced (or re-introduced) some important new ideas into official thinking.[29]

In the Central Scotland White Paper emphasis was on "growth areas chosen as potentially the best locations for industrial expansion" and the building up of these areas "by providing for them, in accordance with a coherent plan, all the 'infrastructure services'..." (Scottish Development Department, 1963, p 5). These proposed growth areas were to be given preferential treatment and be allowed to maintain their development area status (and associated benefits) even if local unemployment rates dropped below the set limit, for under existing legislation (the 1960 Act) areas were automatically de-scheduled when unemployment fell below the specified level. This constituted a welcome shift toward a more positive, long-term view of regional policy; it also high-lighted one of the most serious draw-backs to the 1960 Act, a feature eliminated finally in 1966.

Concern was also expressed for the problems of housing supply and transport and communications as constraints on growth and development -- a new, if somewhat overdue, recognition of the larger implications of what had too long been considered simply a local authority/land use problem. There was, however, a continuation of emphasis on housing supply in terms of housing for 'overspill' populations displaced from Glasgow and other older cities; and to some extent the new towns of Central Scotland were still viewed in this light. Yet it was observed that some early new towns such as East Kilbride, though originally designated almost entirely in terms of housing replacement, had in fact become centers of industrial growth and employment expansion. The White Paper incorporated this idea, and of eight proposed 'growth areas' in Central Scotland, five were centered on new towns.

On the other hand, the White Paper contained little discussion about the nature or scale of regional incentives to industry, either considering them outside its scope or else assuming that they would continue in generally the same way.

The North East White Paper was in many ways similar, though its focus was on a 'growth zone' as opposed to the several separate 'growth areas' of the Central Scotland paper. In reality, however, this was only a vague form of the concept, for the single 'growth zone' included all of the Tyneside and Teeside built-up areas and the whole eastern half of County Durham lying between them. Practically the only areas not included were either rural or isolated from the main industrial centers. Also, the idea of using new towns for regional economic, as well as local housing purposes, was more hesitantly recognized in the North East White Paper than in the Central Scotland one. As for infrastructure, the North East paper's emphasis was primarily on roads and transportation. In terms of existing regional policies, suggestions were made that support should be extended to non-industrial development, especially offices.

Both reports have been criticized, sometimes severely, for being too general to provide guidelines for policy. Yet many of their ideas have since had substantial influence on official thinking despite their failure to immediately lead to legislation. In particular, these White Papers heavily emphasized physical and location factors along with their main concern for economic growth, signalling a new awareness of the interrelationships between economic and physical planning. This awareness, while perhaps more well accepted in some levels of the civil service (and especially the Scottish Office) than publicly indicated, was put forward rather tentatively, for at that time such ideas ran directly counter to the conventional orthodoxy which still rigidly compartmentalized the two.[30]

Additional developments in 1963 included the Local Employment and Finance Acts of 1963. Passed largely in response to administrative difficulties in the 1960 legislation, these new acts simplified the grant system (making it somewhat more generous in many cases) and introduced accelerated depreciation as a new incentive. But the changes were mostly "touching-up" measures rather than real shifts in approach. Also in 1963 the Government set up the Location of Offices Bureau (LOB) to work toward reducing the concentration of office building and employment in London. But in its initial form it was advisory only -- there was no prohibitory or control legislation analogous to the I.D.C. system -- and so LOB could only advise, assist, and persuade; its accomplishments were accordingly minimal. Finally, there were the modest beginnings of a labor retraining scheme, with the 1963 Budget setting aside about $28 million for a program of several years duration.

These activities of 1963 were the last in the period of gradual change, however, for the 1964 General Election ended thirteen years of Conservative rule and brought to power the second post-war Labour Government.[31] National economic planning and regional development had both been made issues in the campaign by the Labour Party's charge that the Conservative Government had failed to institute effective planning or control of the economy. After the victory one of its campaign pledges was quickly redeemed by the Government's establishment of the Department of Economic

Affairs (DEA), a new ministry clearly intended to play the leading role in economic planning and management.[32] As an indication of the importance which was attached to regional considerations, the D.E.A.'s initial organization included a regional planning group as one of its four main internal divisions, thus establishing regional planning at the heart of the new Government's economic policy machinery.

Steps were soon taken to promote a systematic regionalization of operations within the existing governmental institutions. The Government began to establish an officially uniform set of regions, both to help co-ordinate the regional work of the various ministries and departments and also to provide the areal basis for the new regional planning organizations.[33] In each of the new regions there was to be established a Regional Economic Planning Council and a Regional Economic Planning Board (Peterson, 1966; Turnbull, 1967; C.O.I., 1968). The Councils, appointed by the Government, draw upon "leaders" in the region (businessmen, trade union officials, academic, other professionals) and provide a body hopefully both knowledgeable about and roughly representative of the region. The functions of the Councils are advisory:

> "The economic planning councils will be concerned with broad strategy on regional development and the best use of the region's resources. Their principal function will be to assist in the formulation of regional plans and to advise on their implementation. They will have no executive powers."
> (McCrone, 1969, p 23)

The Boards are rather different, being collections of civil servants from various branches of government who are already working in or with responsibility for the region, only the chairman coming directly from the D.E.A. The Boards are expected to help co-ordinate the activities of a wide range of public bodies by bringing together their official representatives.[34] Eventually, as official regions become uniformly used, the representatives on the Boards will be concerned with one region only. It takes time to get such a system organized, staffed, and into operation; but by the late Sixties the regions were established (eight in England, one each in Wales, Scotland, and Northern Ireland) and their Boards and Councils were in operation.

The new Government also quickly moved to increase the emphasis on industrial estates and, especially, advance factories. Industrial estates had been used as a tool of policy since before the war; and built-to-order factories were still being erected throughout the post-war period. But the speculative building of standardized advance factories had been stopped in 1948 and not resumed until 1959, even then only on a small scale. The Labour Government accelerated the rate of construction, and advance factories began going up in fairly large numbers all over the Development Areas[35] (E.F.T.A., 1970; McCrone, 1969, pp 141-142).

Among other new measures was an immediate extension of I.D.C. controls to make certificates mandatory for all industrial development in excess of 1000 square feet, down from the previous 5000 square foot limit. In early 1965 prohibitory control was also extended to office development by the Control of Office and Industrial Development Act, thus putting some teeth into the activities of the L.O.B. Initially restricted to London, this control power was later extended to the Birmingham conurbation (in August 1965) and finally to all of the South East, East Anglia, West Midlands, and East Midlands Planning Regions (in July 1966).[36]

The 1964 Industrial Training Act, passed soon after the change of Government, empowered the Minister of Labour to set up Training Boards for different branches of industry and provided additional finance to expand the number of Government Training Centers; four years later there were 42 training centers with 8,900 places. Tied to this was a provision by which firms in Development Areas became eligible for grants toward the costs of employee training and re-training; accordingly, much more money began being spent on training many more workers.

Finally, steps were taken to remedy the crippling lack of regional statistics by establishing programs for developing compatible sets of disaggregated data from all parts of government. The Abstract of Regional Statistics was begun as an annual compendium of such regional data.

Of course, any new Government has its hands full when it first comes to power, especially after so long a period in Opposition; many serious problems other than regional ones were pressing for attention. Nevertheless, it was made clear that regional policy, along with economic management in general, would be a major concern of the new Government. The most significant evidence of this was seen in September 1965 with the publication of The National Plan (Department of Economic Affairs, 1965a). A document with many short-comings -- not surprising considering the speed with which it was prepared and the lack of previous experience in governmental plan-making -- it provided a strong statement of current ideas, intentions, and hopes. As a guide for national economic management, however, The National Plan did not long survive, being "effectively strangled on the 20th July when the Government introduced a number of severe deflationary measures in the hope of ending the series of sterling crises"[37] (Bailey, 1968, p 66).

But even if its national economic relevance was destroyed, many of The National Plan's ideas for component sectors or problem areas did survive to influence later policy. For example, regional planning had been viewed by the plan not simply as large-scale land use planning but as being "concerned with decisions to develop the infrastructure of the economy, and with the location of employment and population" (Bailey, 1968, p 89).

In line with the central theme of The National Plan, regional policy was viewed in terms of its contribution to national economic growth, particularly with respect to utilization of idle resources and expansion of the labor force. But it did constitute an explicit argument that regional policy could serve both national economic growth and regional redistribution objectives.

In October 1965 the areal definition of aided regions was expanded considerably, with 16 areas being added to the list of Development Districts. Besides simply broadening the coverage, this move also made the Districts into continuous geographical areas.

In November 1965 another new step in regional planning was taken with the establishment of the Highlands and Islands Development Board. Responsible for the 7 "crofting" counties of North-Western Scotland -- a unique area with unique problems -- the Board is itself unique in its wide range of activities and powers.[38] Despite a relatively small budget (about $3.65 million in 1968) and vociferous opposition in certain quarters, the Board in its first five years made itself a recognized and widely accepted force in Northern Scotland. As early as 1968 McCrone could say that "the board in its short existence has had a greater impact on the Highlands than any other body ever has" (McCrone, 1969, p 235). It is a new approach, the experience of which could have important implications for future directions in regional development planning.[39]

In January 1966 two more documents appeared, a new expansion plan for Scotland (Scottish Office, 1966) and a White Paper on Investment Incentives (D.E.A., 1966b). The former, while primarily an economic plan, devoted considerable attention to physical planning, transportation and communication, and housing. The problems of housing indeed loomed large in the analysis, much of the discussion of infrastructure being related more to housing redevelopment and overspill schemes than to directly stimulating growth.[40] But physical infrastructure needs were also discussed at length, and the idea of concentration -- of economic activity, services, and people -- was strongly urged as a way to maximize the growth impact of new development. This growth center idea ran through the study, elaborating the related notions first aired in the 1963 White Paper. Scotland was also analyzed in terms of its constitutent sub-regions, a useful step toward integration with lower level planning.

In late March 1966 another General Election was held, and the Labour Government changed its razor-thin 1964 majority into a commanding lead: 97 seats over all other parties combined and 110 more than the Conservatives. Reinforced by this electoral endorsement of its actions and proposals the Labour Government proceeded to further develop and carry through its legislative programs, generally along the lines of its first 17 months' activities and ideas.

The White Paper on Investment Incentives was followed in August by the Industrial Development Act of 1966, which embodied most of its proposals. The new legislation, in keeping with its central purpose of stimulating investment, instituted a system of 20 per cent Investment Grants for all manufacturing industries throughout the country. A regional differential was introduced by raising this grant to 40 per cent for industries in the Development Areas.[41] (Later these figures were raised to 25 and 45 per cent respectively) The Act also reconstituted the geographical basis of regional policy, replacing the Development Districts with larger and more permanently defined Development Areas. Of particular significance was the changed basis upon which these policy regions were to be defined: whereas the old Development Districts had been created and abolished on the basis of unemployment exclusively, the new Development

Areas were to be selected according to a number of criteria including unemployment, population change, migration, and even 'the objectives of regional policy'. When first set up, there were five of these Development Areas, encompassing in total about 40 per cent of Britain's land area and 20 per cent of its population.[42]

The 1966 Act also reconstituted the basis and mechanism of financial assistance, sweeping away the complex system of negotiable grants and tax incentives which had been in use under the 1963 legislation.

> "The new policy was to rely almost entirely on grants both for the general encouragement of investments and for the assistance of the less prosperous areas. Investment allowances disappeared completely and accelerated depreciation survived mainly in the form of a 15 per cent initial allowance for new industrial building in all parts of the country. ... On buildings (industrial and non-industrial) the basic grant is to remain unchanged at 25 per cent in the development areas. ... Elsewhere in Great Britain no grants are available towards the cost of the buildings. (Wilson, 1967, p 8)

Previously, tax concessions had been set against profits; the argument behind the new system was that such assistance was of little help for new firms in their first few years -- the critical period for new entrepreneurs.[43] Also, the older methods had relied heavily upon screening applications through BOTAC (Board of Trade Advisory Committee), a procedure which theoretically permitted flexibility and adaptability in operation. However, it could equally become a slow and cumberson process, involving a considerable time lag by demanding an elaboration of detailed proposals by applicant firms, and retarding corporate planning through the non-predictability of the level and type of assistance to be made available in any particular case. The new blanket grants greatly simplified matters -- though in practice there was still a serious time lag between incurring a capital expense and receiving the grant. Set against this gain of simplicity, however, is the loss of the old system's flexibility (even though under the new system some discretionary additional help could still be negotiated from BOTAC). Moreover, the new program's assistance was only loosely, if at all, related to the commercial performance of the firm; grants could thus have the effect of artificially shoring up economic failures. In addition, aid was no longer related to the job-producing results of the firm's activities, leaving the way open for a situation in which heavy government assistance might produce very few jobs because of grants going disproportionately to capital-intensive industries.[44] However, the extra assistance negotiated through BOTAC was still related to job creation.

In April 1967 the Government produced a "Green Paper" on the Development Areas in which a Regional Employment Premium (REP) was proposed (D.E.A., 1967; Brown, Lind & Bowers, 1967). The scheme, put into effect in September 1967, involved payment of $3.60 per week per adult male employee (lower rates for women and juveniles) to manufacturing industries in the Development Areas. These industries were also allowed to retain a premium which had earlier been added to the regular rebate of Selective Employment Tax (SET) which had been abolished in 1967 for employers outside the Development Areas. These two measures together provided a subsidy of $4.50 per week per man -- and this considerable amount was not dependent upon profit performance, creation of new employment, or any other such measure. This was unquestionably a huge subsidy, both in absolute terms (expected to be about $240 million in a full year) and in terms of its value to the recipient firm (McCrone estimated it to represent something like an 8 per cent reduction in an average firm's wages bill). Moreover, it was a clear departure from past policy in that it was the first large-scale subsidy to operating costs -- other measures usually concentrated on initial or capital costs. One economic argument put forward in its favor was that such a selective labor subsidy represented a quasi regional devaluation and that it thus provided the poorer regions with a labor cost advantage otherwise unavailable in a country of nation-wide wage bargaining.[45] Another important justification of REP on such as scale was that a labor subsidy was needed to "balance" the capital grants given under the 1966 Act and thus avoid any undue bias toward capital-intensive industry. In 1970 the SET premium was withdrawn, reducing the subsidy back to the REP level of $3.60 per man. Interestingly, this withdrawal was described as the means by which funds would be raised to finance the "grey area" provisions of the 1970 Local Employment Act.

In May the Agriculture Act of 1967 was passed, incorporating into its Part III a new feature:

> "...for meeting the special problems of the development as rural areas of hills and uplands, and the special needs of such areas, the appropriate Minister may...establish a Board, to be known as a Rural Development Board, for any area appearing to be one where these special problems or needs exist." (Agriculture Act, 1967, §45(1))

First suggested in an August 1965 White Paper on "The Development of Agriculture", such boards were intended to provide the medium for redevelopment and reorganization of relatively marginal (mostly hill-country) rural areas. Their principal concerns were to be with agricultural and forestry land uses, and with the fostering of such activity through farm amalgamation, financial assistance, and general support and coordination. Their main tools were to be the power to control land transfers, to control afforestation, to buy (through agreement usually, by compulsory purchase only rarely) and hold and manage land, to provide finance for necessary services and certain other functions, to elaborate "a coordinated scheme of amalgamation of agricultural land, reshaping of agricultural units, and afforestation", and to coordinate associated developmental activities in such rural areas. In June 1967 the Government announced their intent to set up such a board for the mountainous area of North Central England, and after public hearings, Parliamentary passage, and other procedures, the Northern Pennines Rural Development Board assumed its full powers on 1st November 1969. In the meantime an effort was made to set up such a board for Mid-Wales, but the inept handling of the attempt aroused furious opposition (through exaggerated fears of forced amalgamation) and forced the Government to give up the attempt.

A by-product of the earlier REP legislation was the setting up, in September 1967, of the Hunt Committee. A number of regions which did not qualify as Development Areas but which nonetheless could not be considered "prosperous" had for some time felt themselves 'left out' by regional policy. This applied particularly to areas immediately adjacent to the designated Development Areas. Stimulated by the $240 million per annum that REP would be pumping into the Development Areas, these regions (later called the "grey areas") strongly agitated for some consideration of their own problems and needs. The Government agreed to examine the problems of such 'intermediate areas' and accordingly set up a committee

> "to examine in relation to the economic welfare of the country as a whole and the needs of the development areas, the situation in other areas where the rate of economic growth gives cause (or may give cause) for concern, and to suggest whether revised policies to influence economic growth in such areas are desirable and, if so, what measures should be adopted."
>
> (D.E.A., 1969a, p 1)

This Committee of Inquiry into the Intermediate Areas (commonly called the "Hunt Committee" after its chairman) reported in April 1969.

In November 1967 came the designation of Special Development Areas: small regions within the existing Development Areas which, because of some special economic distress, were made eligible for extra assistance in addition to regular Development Area aids. Initially, and until 1971,

Special Development Areas were almost entirely coal mining regions, particularly those hit hard by accelerating pit closures.[46] Special Development Area assistance, extended usually to new manufacturing projects, included rent-free periods of up to 5 years in industrial estate premises (2 year in Development Areas), building grants of 35% (normally 25% in Development Areas), loans toward the balance of building costs, and special 'operational grants' to help in the early years of establishment (negotiable, and related to employment creation). Clearly a special welfare provision, this new measure ran counter to current ideas of concentration and "growth centers", for it put even more money into exactly those places with arguably the least discernible potential for growth.[47] It is fair to say, on the other hand, that the mining areas represented special problems needing special treatment, even if only as a short-term measure during the worst period of run-down.[48]

In early 1969 "The Task Ahead: Economic Assessment to 1972" (D.E.A., 1969b) was published as an updated statement of the Government's economic aims. For the nation as a whole certain "broad objectives" were laid down as the basis for obtaining a "satisfactory rate of growth of output":

> "First, the achievement of a substantial surplus in the balance of payments...
> Secondly (and partly to maintain the better trade balance) a steady improvement of the competitive efficiency of the economy...
> Third, a fuller utilisation of resources, including an improvement in the regional balance of the economy; this involves a further and more marked reduction in the disparity between rates of unemployment in the Development Areas and the rest of the country." (D.E.A., 1969b, p 6)

Regional "balance" was once more confirmed in its place of importance among the nation's economic objectives -- the place it had held officially since the White Paper on Employment Policy twenty-five years earlier. "The Task Ahead" chapter on 'Regional Strategy and Prospect' offered little in the way of new suggestions or programs; the Government apparently felt that most of the necessary tools of regional policy were in existence and it only remained to vigorously apply them. Also interesting is the document's emphasis on the role of the Regional Economic Planning Councils and Boards in the formulation of regional policies and strategies -- perhaps a suggestion of regional decentralization.

In April 1969 the "Hunt Committee" Report on the problems of 'Intermediate Areas' was published (D.E.A., 1969a). The Committee identified those areas of concern under their terms of reference to be, mostly, the older industrial regions of the North West and Yorkshire and Humberside, though certain other smaller areas were also included. In evaluating the nature of the problem and the possibilities of solution, the Committee emphasized the "Total Environment" as a factor:

> "We mean by this the full range of the social, educational, cultural, industrial, and commercial facilities as well as the physical infrastructure of buildings, roads, docks, and the like which houses and sustains them. Our conclusion is that poverty in the total environment is often associated with slow economic growth and net outward migration, and is an important component of the complex of inter-acting factors which make it difficult for areas to recapture their former dynamism." (D.E.A., 1969a, para 29)

Thus when actual recommendations were made concerning inducements for industry the Committee warned:

> "The industrial renewal which our recommendations are intended to foster cannot, however, be separated from its physical and social setting. The improvement of the environment of these areas is just as important in the longer term as direct measures for industrial modernisation." (D.E.A., 1969a, para 372)

Their specific policy recommendations included: (1) 25% grants for industrial building; (2) extension of the Development Area training measures to the Intermediate Areas; (3) extension of the activities of the Industrial Estates Corporation into these areas; (4) a new and accelerated program of derelict land clearance, including extension of the Development Area policy of 85% assistance to local authorities; (5) establishment of "growth zones";

(6) raising the I.D.C. limit to 10,000 square feet and making them more freely available within the Intermediate Areas; and (7) de-scheduling of Merseyside, i.e., downgrading it from Development Area to Intermediate Area. It was intended that such measures should be extended to most, if not all, of the North West and Yorkshire and Humberside regions. In his note of dissent, however, Prof. A. J. Brown questioned these main ideas. He thought too much emphasis had been put on infrastructure as an incentive to industrial growth rather than on financial and fiscal measures. He questioned the wisdom of raising the I.D.C. limit or relaxing the controls because of the danger to the Development Areas. Finally, he doubted that Merseyside was ready for such a drastic step as de-scheduling.

In any case, the Report quickly became a "dead letter", as Odber observed:

> "...government reaction came in a matter of days, but clearly neither the Hunt Report itself nor Brown's 'minority report' is being implemented..." (Odber, 1970, p 205)

It is uncertain exactly why the Government's reaction was so meagre. Odber suggests that one fundamental problem was

> "...the absence of any good techniques of measuring the impact of the different policy weapons. This difficulty stands out as crucial throughout the Hunt Report and leads to directly opposed recommendations in the Report and the note of dissent." (Odber, 1970, pp 205-6)

While no one can contradict Odber about the lack of good analytical technique, there may have been other reasons for the Government's inaction. An obvious first suggestion is financial: given the mushrooming cost of existing regional policies and the continuing crisis in the national economy, the Government may have simply been unwilling (or unable) to countenance any major new programs. The Government may also have felt that an extensive Intermediate Area program would necessarily detract from growth in the Development Areas and that the latter should retain clear priority.

In the end, the official response was the Local Employment Act 1970, which fell far short of the Hunt Committee recommendations. It designated a number of small Intermediate Areas which encompassed only a minor proportion of the whole North West and Yorkshire and Humberside regions. It also added a few other small areas, in the South West, South Wales, Scotland, and the East Midlands - the first three being "transitional" areas adjacent to Development Areas and the last being a problem coal field area. The Government also declined to de-schedule Merseyside. The Act did include a 25% building grant, but of course it applied only in the small areas designated. There was not any special emphasis on growth zones; indeed, the officially chosen Intermediate Areas tended to be the most difficult areas in which to stimulate industry — the reverse of a growth zone policy. The Act did, however, provide for industrial estate factories and the full range of training assistance, all on the same basis as the Development Areas. Finally, the Intermediate Areas were given a 75% grant for derelict land clearance — and this one measure was extended to the whole of the North West and Yorkshire and Humberside, as well as to parts of the north Midlands.

The Reports of two Royal Commissions on Local Government Reform appeared in 1969, in June for England (Royal Commission on Local Government in England, 1969) and in September for Scotland (Royal Commission on Local Government in Scotland, 1969). They are of interest not only because of the importance that both attached to planning generally but also because each proposed some form of regional or provincial level of government. In both England and Scotland the existing form of local government dates from the Victorian era and is widely acknowledged to be ill-fitted to the needs of modern life. Typical failings of the systems were identified as fragmentation of authority, division of responsibility, too-strict separation of city from country, inadequate size of authorities for supporting effective services, and the lack of both metropolitan and provincial levels of government. Given the different settlement patterns, population densities, and traditions of the two countries, the two Commissions took different paths in seeking to rectify these failings.

One of the major recommendations of the Redcliffe-Maud Commission (England) concerned the need for provincial councils:

> "Our examination of England had also led us to a second firm conclusion: local government, however organised, needs to include a new representative institution with authority over areas larger than any city region, not unlike the eight areas of the present regional economic planning councils. This 'provincial council' would handle the broader planning issues, work out provincial economic strategy in collaboration with Central Government and be able to act on behalf of the whole province." (Royal Commission on Local Government in England, 1969 (short version), p 7)

In fact the Commission proposed eight such provincial councils, with boundaries quite similar to those of the Economic Planning Regions. At the lower levels, 61 unitary authorities were proposed, three of which were to be metropolitan authorities for the Birmingham, Manchester, and Liverpool areas. Except in the three metropolitan areas (and London, which was outside the Commission's terms of reference) there would be only one effective level of government, for the provincial councils would be elected by the local councils and have only the broad responsibilities outlined above.

In contrast, the Wheatley Commission suggested a "2-tier" system for Scotland, with 7 regions and 37 districts -- these regions to have "major" governmental powers. There was not proposed a metropolitan authority for the Glasgow area, but the West Central Scotland region would tend to function to some extent as such a body. None of this would interfere with the existence of the Scottish Office and its central role as economic planning body for Scotland as a whole.

Though the Labour Government accepted the Commission reports in principle, no specific legislative proposals for local government reform were brought forward by them.

The last Labour measure affecting regional policy was the governmental reorganization of October 1969. The principal feature of this rationalization was the creation of two new Whitehall "empires", each bringing together under one over-all head a number of departments or functions formerly independent. At the same time the Department of Economic Affairs was formally ended and its responsibilities parcelled out.[49] One of the new groupings brought together the Ministry of Housing and Local Government and the Ministry of Transport, along with most of the regional activities from the expired D.E.A., under the Secretary of State for Local Government and Regional Planning. The other major grouping was an expansion of the Ministry of Technology to encompass the Ministry of Power and the industrial location activities which were taken away from the Board of Trade. Clearly an improvement in administrative structure, the reorganization did leave two apparent anomolies: first, regional development and planning remained separate from location of industry activities; and second, the Secretary of State for Local Government and Regional Planning in fact secured responsibility only for England, with the Welsh and Scottish Offices remaining autonomous in these fields.

As Britain entered the 1970s, then, its body of regional policy was about as extensive as an enthusiastic regionalist could have hoped for a decade earlier. For regional policy purposes the country was divided into four types of area:

1. Development Areas, large regions in which a wide range of incentives and assistance was available;

2. Special Development Areas, small areas within the regular Development Areas for which the ordinary Development Area benefits were supplemented by a number of extra aids.

3. Intermediate Areas, mostly small areas wherein a reduced scale of assistance was available.

4. The Rest of the Country, where no special inducements or aids were available -- but for parts of which certain negative controls and restraints might be applicable.

The various tools of policy could be categorized as: first, direct government assistance to enterprises; second, indirect assistance to promote regional development generally; and third, negative controls to shift growth away from the prosperous areas to the assisted areas.

An idea of the range of these programs is given by a simple listing of the different aids available in mid-1970 in the Development Areas alone (and also available in the Special Development Areas):

I. Direct central government assistance to firms expanding in or moving into the Development Areas:

1. Grants of 40 % (instead of the nation-wide standard 20%) for investment in plant and machinery.

2. Grants of 25 to 35 % for industrial buildings.

3. Loans, negotiable from MOTAC (successor to BOTAC), for expenditures on plant, premises, or working capital.

4. Availability of factory premises on industrial estates, either standard or custom-built, in some cases with rent-free periods of tenancy or special terms of purchase.

5. Grants through MOTAC toward expenses of moving.

6. Grants, for manufacturing industries, on a per-worker basis through the Regional Employment Premium.

7. Numerous forms of assistance for government and in-house labor training and re-training.

8. Assistance for transfer of key workers.

9. In certain cases, assistance for tourist industries.

II. Indirect government assistance to boost development generally:

1. Special assistance can be given for provision of "basic" services where this will contribute to the development of industry.

2. Local authorities may make use of a higher quota of borrowing from the Public Works Loan Board.

3. Local authorities are allowed to lend money to industrialists at attractive rates of interest for factory building.

4. Local authorities are encouraged to clear derelict land through a system of 85% grants from central government.

5. Where practicable, government purchasing authorities can give preference in contracts to firms in Development Areas.

6. In selected cases, the Electricity Generating Boards may negotiate special contracts for long-term electricity supply.[50]

7. Central government may allocate a 'larger than fair share' amount of infrastructure investment capital to the Development Areas.

III. Measures to increase the supply of new development by controlling growth in the more prosperous regions:

1. Industrial Development Certificates are required for new or expanding industrial buildings over 3000 sq ft in the Midlands and South East and over 5000 sq. ft. elsewhere in the country; certificates are usually freely given in the Development Areas but given in other areas only where it is not possible to expand in the Development Areas.

2. Office Development Permits are required for new office building in the South East and Midlands. Efforts are made to induce office development to go to the Development Areas, where no permits are required.

There were slightly different sets of policies available in the other types of assisted areas.[51] There were also special sources of extra assistance such as the Highlands and Islands Development Board. Most significant, however, was simply the immense amount of money being spent: by 1968/69, when most of the above measures had come into force, about $730 million was being spent annually on regional policy.[52] Certainly, in reaching its "new era" British regional policy had come a long way from the Thirties.

The Future of Regional Policy: 1970-1971

On 18 June 1970 a General Election turned Labour out of office after five years and eight months.[53] By bringing in a Conservative Government which had in Opposition fought many of the recent regional policy measures -- and which was ideologically committed to reducing government activity generally -- the election seemingly marked an end to the "New Era" of regional policy.

Changes began soon after the new Government took office. Administrative steps were taken to relax office and industrial development controls, both of which are disliked in Conservative circles. The future of the Highlands and Islands Development Board also came into question, as its Chairman and members ended their five-year terms of office just in time for their successors to be named by the Conservative Government. Its new Chairman is a retired diplomat whose lack of direct qualifications baffled even The Times, while the Board acquired enough conservative members to suggest that the Government intends to "tame" the Board.[54] The Government also satisfied a long-standing clamor within the Conservative Party by announcing that the Regional Employment Premium would be discontinued in September 1974 (the earliest date which would still honor the previous Government's original establishment of REP as a seven year program). In addition, the Government announced its intention not only to decline to establish new Rural Development Boards but also to close down the one which had actually got going; and in March 1971 the Northern Pennines Rural Development Board was closed.[55] A few other new measures were also ended (or allowed to expire), such as a hotel grant scheme which was finished up in March 1971.

The Government fulfilled another campaign pledge by announcing the replacement of investment grants with tax allowances; in an October White Paper on "Investment Incentives" (Department of Trade and Industry, 1970) and in other official releases at that time, specific details of the new policies were given. No grants would be paid for capital investment expenditures incurred after 26 October 1970, anywhere in the country. To provide a general incentive in place of grants, Corporation Tax was reduced by 2½ percentage points and legislation was introduced for a new system of early depreciation allowances for expenditure on plant and machinery.[56] To provide an element of differential regional assistance there will be free depreciation on plant and machinery expenditures in the Development Areas. In addition, the temporarily increased rate of initial allowance of 40% for industrial buildings in Development and Intermediate Areas will be continued indefinitely -- though the 30% rate now applicable in the rest of the country will revert to 15% in 1972. Also, the rates of building grants for Development Areas are raised from 25% and 35% up to 35% and 45%. The Intermediate Areas do not share in this raised rate, but they do benefit from a relaxation of the "cost per job" limitations.[57] Finally, one of the principal extra incentives available in the Special Development Areas was changed. Previously, the "operational grants" had been calculated at a rate of 10% of cumulative costs of new buildings and plant and machinery in the first three years of the project's operation; the new basis was 20% of wage and salary costs of the project during its first three years, subject (as before) to an upper limit relating to jobs provided.

The Government also announced their intent to increase the assistance provided under the various Local Employment Acts, as promised in their campaign, in order to put more emphasis on the "selective" instruments of regional policy. The Labour Government, however, had already made a dramatic increase in Local Employment Act spending, raising it from $104.5 million in 1967/68 to $201.4 million in 1969/70. So even if the new Government increases spending by the $60 million (£25 million) promised in their White Paper it will be rather less of an increase than that made by their predecessors.

Along these same lines the Government emphasized, both before and after the election, their intent to rely on increased infrastructure spending as a key to regional development. But again, there had already been a massive build-up of such infrastructure spending in the assisted regions during Labour's term in office. According to one writer:

> "The very sketchy data for the pre-Hailsham years -- about 1960-62 -- suggest that such spending was about in proportion to population (the present development areas have about 22% of the UK population). By 1966, their share of this spending had risen to over 30%, and for the last financial year I estimate it at 37%." (Malcolm Crawford in the Sunday Times, 21 February 1971)

As measured by public investment in new construction, for example, the four poorer regions in 1968-69 received an average of £53.40 of new construction per capita, as opposed to a U.K. average of £42.30; just three years earlier, in 1965-66, the poorer regions' average was £35.80 and the U.K. £30.30. On this basis the amount of infrastructure spending per capita in the four poorer regions, already 18 per cent above the national average in 1965-66, rose by almost 50 per cent in three years, ending up over 25 per cent above the national average.[58] All of this makes it difficult to see how the present Government will be able to increase, or indeed even maintain, these differential levels of infrastructure expenditure.[59]

During these first months of the new Government the economic situation generally remained about as expected and most of the early policy measures fairly closely followed previously formulated intentions. Toward the end of 1970, however, the economic situation began to deteriorate sharply; the third and then the fourth quarters each showed significant decline of activity, while the first quarter 1971 brought the highest figures for over thirty years.[60] The poorer regions, already weak from four years of economic "squeeze" and slow growth, suffered particularly.

TABLE 3

UNEMPLOYMENT RATES IN THE REGIONS
(AVERAGES OF MONTHLY FIGURES, PER CENTS)[61]

<table>
<tr><th></th><th>1960</th><th>1961</th><th>1962</th><th>1963</th><th>1964</th><th>1965</th><th>1966</th><th>1967</th><th>1968</th><th>1969</th><th>1970</th><th>1971*</th></tr>
<tr><td>South East</td><td rowspan="2">1.0</td><td rowspan="2">1.0</td><td rowspan="2">1.3</td><td rowspan="2">1.6</td><td rowspan="2">1.0</td><td>0.9</td><td>1.0</td><td>1.7</td><td>1.6</td><td>1.6</td><td>1.6</td><td>1.7</td></tr>
<tr><td>East Anglia</td><td>1.3</td><td>1.4</td><td>2.1</td><td>2.0</td><td>1.9</td><td>2.2</td><td>2.7</td></tr>
<tr><td>South West</td><td>1.7</td><td>1.4</td><td>1.7</td><td>2.1</td><td>1.5</td><td>1.6</td><td>1.8</td><td>2.5</td><td>2.5</td><td>2.7</td><td>2.9</td><td>3.0</td></tr>
<tr><td>East Midlands</td><td rowspan="3">1.0</td><td rowspan="3">1.1</td><td rowspan="3">1.6</td><td rowspan="3">2.0</td><td rowspan="3">1.0</td><td>0.9</td><td>1.1</td><td>1.8</td><td>1.9</td><td>2.0</td><td>2.3</td><td>2.5</td></tr>
<tr><td>West Midlands</td><td>0.9</td><td>1.3</td><td>2.5</td><td>2.2</td><td>2.0</td><td>2.3</td><td>2.4</td></tr>
<tr><td>Yorkshire & Humberside</td><td>1.1</td><td>1.2</td><td>2.1</td><td>2.6</td><td>2.6</td><td>2.9</td><td>3.1</td></tr>
<tr><td>North West</td><td>1.9</td><td>1.6</td><td>2.5</td><td>3.1</td><td>2.1</td><td>1.6</td><td>1.5</td><td>2.5</td><td>2.5</td><td>2.5</td><td>2.7</td><td>3.1</td></tr>
<tr><td>North</td><td>2.9</td><td>2.5</td><td>3.7</td><td>5.0</td><td>3.3</td><td>2.6</td><td>2.6</td><td>4.0</td><td>4.7</td><td>4.8</td><td>4.8</td><td>4.9</td></tr>
<tr><td>Wales</td><td>2.7</td><td>2.6</td><td>3.1</td><td>3.6</td><td>2.6</td><td>2.6</td><td>2.9</td><td>4.1</td><td>4.0</td><td>4.1</td><td>4.0</td><td>4.1</td></tr>
<tr><td>Scotland</td><td>3.6</td><td>3.1</td><td>3.8</td><td>4.8</td><td>3.6</td><td>3.0</td><td>2.9</td><td>3.9</td><td>3.8</td><td>3.7</td><td>4.3</td><td>5.0</td></tr>
<tr><td>Northern Ireland</td><td>6.7</td><td>7.5</td><td>7.5</td><td>7.9</td><td>6.6</td><td>6.1</td><td>6.1</td><td>7.7</td><td>7.2</td><td>7.3</td><td>7.0</td><td>7.0</td></tr>
<tr><td>UNITED KINGDOM</td><td>1.7</td><td>1.6</td><td>2.1</td><td>2.6</td><td>1.7</td><td>1.5</td><td>1.6</td><td>2.5</td><td>2.5</td><td>2.5</td><td>2.7</td><td>2.9</td></tr>
</table>

(* = First 4 months only, seasonally adjusted)

In Scotland, for example, the unemployment rate went from 4.0% in the second quarter 1970, to 4.3% in the third, 4.6% in the fourth, and approximately 5.2% in the first quarter 1971 -- despite one of the mildest winters in memory. And the losses from the Rolls-Royce failure had not yet been felt! In such circumstances the Government has had little choice but to abandon some of their prior notions about regional policy and try to bolster the regional economies and soften the growing impact of unemployment.

The Government's response in February 1971 was as dramatic as unexpected: several sizeable regions were designated as Special Development Areas, and the incentives applicable to them were increased. A large part of West-Central Scotland, including the whole of Clydeside (Glasgow area), was so designated, as was a large part of North-East England, including Tyneside (Newcastle area) and Wearside (Sunderland area); in addition, parts of the South-East Wales mining region which had not been made Special Development Areas in 1967 were designated.[62] Formerly, the Special Development Areas were mostly mining locales, not large industrial centers, and until February 1971 they included only 1.8% of the insured population of Great Britain. With the inclusion of two conurbations and other urban centers, however, the Special Development Areas now include 8.5% of the insured population. At the same time, the "operational grant" for these new Special Development Areas was raised from 20% to 30% of the first three years' wage and salary costs. Combined with the 45% building grant and the possibility of five years' rent-free tenancy in an industrial estate factory, this substantial payroll subsidy means that the new Special Development Areas are extremely attractive to incoming industry and possess a considerable edge over the unassisted parts of the country and over the regular Development and Intermediate Areas.

Shortly afterwards, in March 1971, a number of places were added to the category of Intermediate Area: Edinburgh and vicinity, a small adjacent to the South West Development Area, a small area in the West Midlands along the Welsh border, and a small area on the Yorkshire coast.

The full amount of extra expenditure all this will involve is uncertain, depending as it does on the response of private industry. But it is given as a rough estimate that the Special Development Area designations will involve about £25 million ($60 million) per annum in additional funds. Together with the £25 million promised in October 1970 (in the form of higher grants on new industrial buildings and certain local government works) this will more or less compensate for the £50 million ($120 million) which the assisted regions lost, according to Government estimates, when investment grants were dropped. The net result of the changes, then, has been to redistribute roughly the same sum, with less for the Development Areas as a whole but more for the Special Development Areas. This is perfectly consistent with the Government's stated desire for "selectivity" and "concentration", as is its previously announced intention to rely more heavily upon the Local Employment Acts while doing away with some of the blanket grants. On the other hand, since REP is still being paid until 1974 the total sum now being spent on regional policy is probably roughly equal to that before the change of Government.[63] This, of course, runs directly counter to the Government's general desire to "disengage" from what they consider excessive governmental involvement in the economy.[64] Even so, it was clear by March 1971 that regional development was still very much a fundamental part of Britain's social and economic policy, notwithstanding the change from a Labour to a Conservative Government. Indeed, following these Spring 1971 announcements, the Secretary for Trade and Industry declared that the Government was not planning any further changes in regional industrial policy; the "New Era" of regional policy looks like continuing for some while yet.[65]

There had been developments under the new Government in the realm of governmental organization which also affected regional policy. First, in October 1970, there was a reorganization of central government. A single "super-ministry" was established — the Department of the Environment -- headed by a Minister possessing not only the statutory powers previously wielded by the separate ministers but also firm control over all departmental finance.[66] The new Department oversees land use planning, housing, transport and transport planning, preservation of amenity, countryside protection, water supply and sewerage, control of pollution, and local government -- as well as a number of other functions less directly related to planning. It also has the "leading responsibility" for regional policy, with important executive powers for the development of regional infrastructure and the maintenance of regional services. Regional industrial policy, however, remains with the Department of Trade and Industry (successor to both Board of Trade and the Ministry of Technology). The Scottish and Welsh Offices, moreover, maintain separate jurisdiction over many of the various aspects, the Department of Environment having its full authority only over England. Thus, while the reorganization carries further the rationalization begun a year earlier under Labour and provides a better basis for coordinated urban and countryside planning, it retains the familiar split between regional planning and regional industrial activities.

There was a second flurry of activity in February 1971 with the publication of three White Papers on local government reorganization, one each for England (Department of the Environment, 1971), Wales (Welsh Office, 1971), and Scotland (Scottish Office, 1971). There were few surprises, however, for as expected the Scottish proposals closely followed the Wheatley Commission's recommendations while the English and Welsh schemes came out quite different from what had been suggested in Redcliffe-Maud.

The Scottish system is to be 2-tier, with 8 regions and 51 districts.[67] The regional authority functions will encompass "major planning and related services" including strategic planning, industrial development, transportation, countryside and tourism, and water and sewerage, as well as "regional housing". The district authorities will have local planning and associated services, building control, housing, and local environmental health. There is not to be a metropolitan government for Central Clydeside, though the 7 new districts in the area (including an enlarged city of Glasgow) will provide something of this, especially with the unifying influence of a single regional authority over them. Such a system, when combined with the strong all-Scotland organization already existing in the Scottish Office, should provide the basis for improved and coordinated planning at all levels.

On the other hand, the Government's proposals for England bear only a partial resemblence to the Redcliffe-Maud Commission's proposals; on the whole, they constitute "tinkering" more than fundamental reconstruction. The Government dropped any idea of regions or provinces, on the grounds that they must wait for the report of the Crowther Commission (Commission on the Constitution); in the meantime they propose to simply continue the present Regional Economic Planning Councils and Boards. The reorganization will create 44 new "counties" in place of the patchwork of counties, county boroughs, and municipal boroughs now in existence. Six of these will be "metropolitan counties" covering the areas around the five conurbations (Newcastle, Liverpool, Manchester, Leeds-Bradford, and Birmingham) and around Sheffield. These metropolitan counties will be subdivided into districts and the powers divided between them in a way somewhat analogous to the present organization of Greater London and also similar to the original Commission proposals.[68]

The 38 counties covering all the rest of England, however, will not be unitary all-purpose local authorities as in Redcliffe-Maud, but instead only dressed-up versions of the present counties, in most cases. The new county authorities will have power over the whole of their geographic areas, as the present division into mutually exclusive counties and boroughs is to be ended.[69] Districts will be designated within the counties, though they will have rather less power than the metropolitan districts -- and considerably less power than now exercised by county boroughs. Areally, the new counties are generally just the old ones or aggregates of the old ones, historical boundaries being adhered to whenever possible. In the division of powers, housing is relegated to the districts and not given to the large, powerful county authorities; in addition, planning powers are split, with some development control being exercised separately at both levels and some other powers being exercised concurrently. It is a strangely mixed and in some ways unsatisfactory solution that is being put forward for non-metropolitan England.[70]

The Welsh proposals follow the pattern of the English ones, though they are spelled out in more detail. There are to be 7 new counties, including three in the heavily populated South Wales area; there will be 36 districts within them, including 17 in the three South Wales counties. As in England, most of the boundaries follow traditional lines (though the 4 new counties outside South Wales will replace 11 separate historical counties) and the division of powers and functions is apparently to be much the same.

The Government hopes to submit legislation in the 1971/72 session and pursue its implementation so that the new systems could be fully operative by April 1974 in England and Wales and early 1975 in Scotland.

Conclusion

The specific features of British regional policy tend to change frequently -- parts of this paper may well be outdated before it reaches print. But the general features are more stable. Perhaps most remarkable is the degree of acceptance of such active public planning -- acceptance by the various Governments of the day, by the Civil Service generally, by the media, by the academic world, and by the populace at large. The antiquated debate over the propriety of public planning, still quite important in the U.S., is not only irrelevant but almost unheard. Moreover, Britain can now draw on nearly four decades of experience in deliberate regional policy making; the first faltering half-steps toward regional policy in the U.S. were only taken in the Sixties. Finally, the magnitude of British regional policy expenditures (about $2,520 million, or £1,050 million, in the four years 1967/68 through 1970/71) indicates the prominent position -- almost unique in the West -- it occupies among national priorities.[71]

Yet, while the development of regional policy making in Britain may seem very highly advanced, a closer view reveals a less satisfactory picture. Too many serious problems remain unsolved, too many vital questions remain unanswered, to leave British regional policy makers with more than two cheers.[72] For example, there has not been developed, either within government or in the Universities, any reasonably unambiguous or generally accepted method for determining the "worthwhileness" of policy measures, individually or in aggregate. Vast sums of money have been expended, but arguments defending such expenditures still rest largely on assertion (however plausible) or even on faith. For that matter, most Governments have failed to state the objectives of their regional policies, except in the most vague and generalized way.

Moreover, criticism can be levelled against many of the more technical aspects of regional policy. For example, if American programs can be faulted for excessive reliance on "demonstration projects", British programs can be faulted for the reverse: the almost total lack of such experimental or pilot programs and projects. Equally, programs tend to be altered or abolished with little specific objective evaluation, political or ideological criteria often proving more potent determinants of policy change. The rapid succession of very different programs of grants and tax incentives, for instance, illustrates the tendency to jump about in policy making with no effort to proceed in an orderly fashion on the basis of deliberate evaluation and redesign. While this _ad hoc_ sort of "get on with the job" approach may have proved effective and useful in the past, British regional policy at this stage certainly needs more systematic and intelligent control and guidance.

In terms of more specific points, recent regional policy can be criticized on a number of grounds: its myopic concentration on manufacturing industry; its lack of consistent spatial selectivity; its over-concern for capital costs only; its all-too-frequent isolation from national economic policy-making; its chronic isolation from local planning; its lack of attention to social overhead capital; its failure to support urgently-needed ancillary research; and its cavalier attitudes toward the information and statistical data needs of program evaluation and control. Certainly there are understandable reasons for most of these failings; and some may not fairly be considered failings of regional policy _per se_. But they are sore enough points to suggest that, despite the effort and resources that have gone into it in the past, British regional policy remains far from being a "finished product".

Indeed, despite its generally quite creditable record of achievement, regional policy needs a lot more (and a lot faster) creative development before it becomes the coherent and reliable instrument desired by society. But at the very least, the start made over the last several decades provides a base and a momentum which could carry Britain through its regional problems more successfully than any other country in the West.

* * * * *

The author would like to acknowledge the assistance of many colleagues at the University of Glasgow whose helpful comments on drafts and sections were highly valuable: Kevin Allen, Gordon Cameron, Robert Grieve, Ian Logan, John Money, Derek Nicholls, and Peter Smith. The interpretations expressed here, however, are of course solely the responsibility of the author.

FIG. 1
ECONOMIC PLANNING REGIONS
INSET
N
ESTIMATED 1971
POPULATION IN ()
IN MILLIONS
SCOTLAND
(5.20)
NORTHERN
IRELAND
(1.53)
NORTH
(3.36)
YORKSHIRE
&
HUMBERSIDE
(4.83)
NORTH
WEST
(6.81)
EAST
MIDLANDS
(3.40)
WEST
MIDLANDS
(5.21)
EAST
ANGLIA
(1.70)
WALES
(2.74)
SOUTH
EAST
(17.41)
SOUTH WEST
(3.79)
MILES
0 20 40 60 80 100
0 40 80 120 160
KILOMETERS

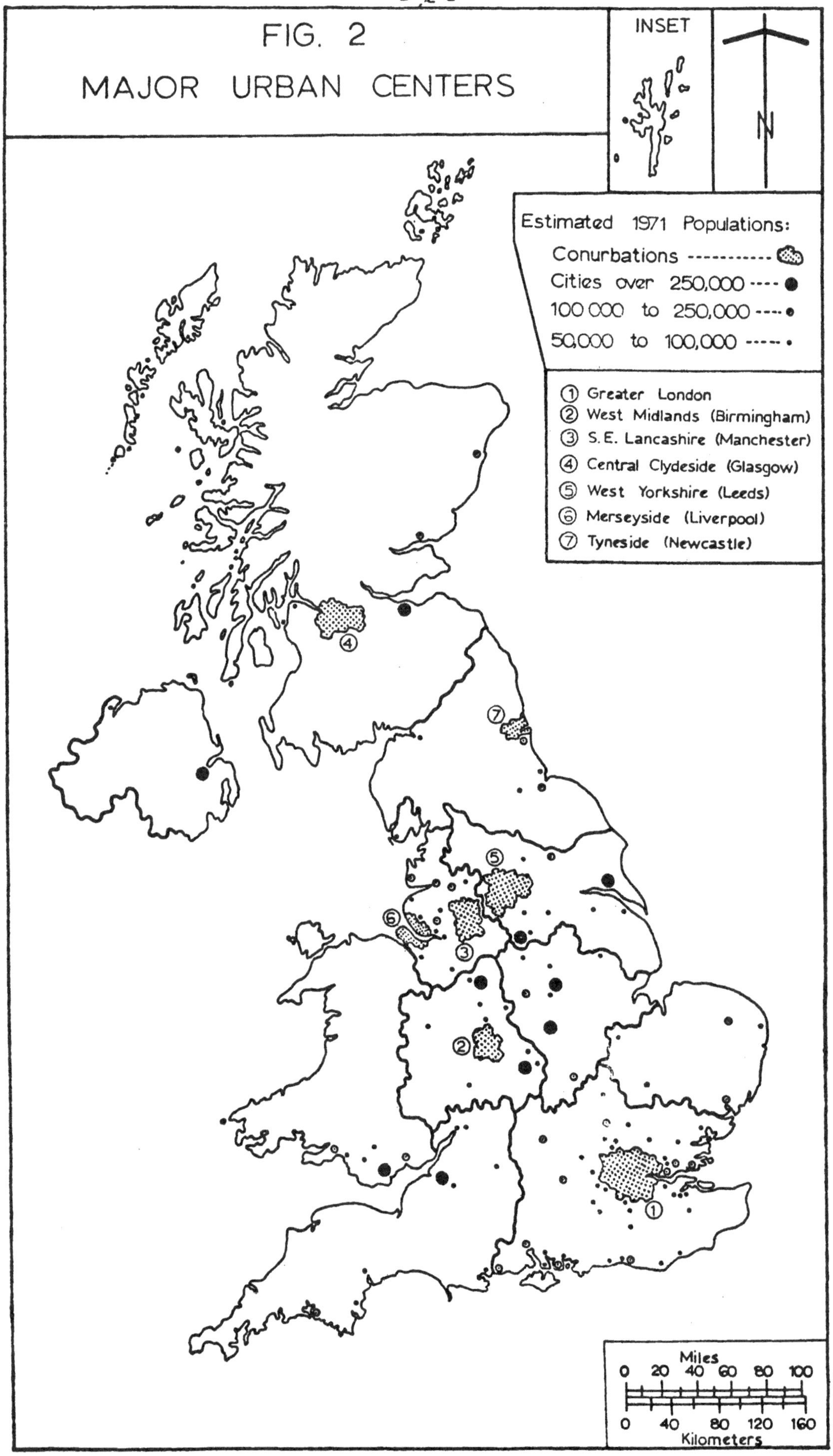

FIG. 2

MAJOR URBAN CENTERS

INSET

N

Estimated 1971 Populations:

Conurbations

Cities over 250,000

100,000 to 250,000

50,000 to 100,000

① Greater London
② West Midlands (Birmingham)
③ S.E. Lancashire (Manchester)
④ Central Clydeside (Glasgow)
⑤ West Yorkshire (Leeds)
⑥ Merseyside (Liverpool)
⑦ Tyneside (Newcastle)

Miles

0 20 40 60 80 100

0 40 80 120 160

Kilometers

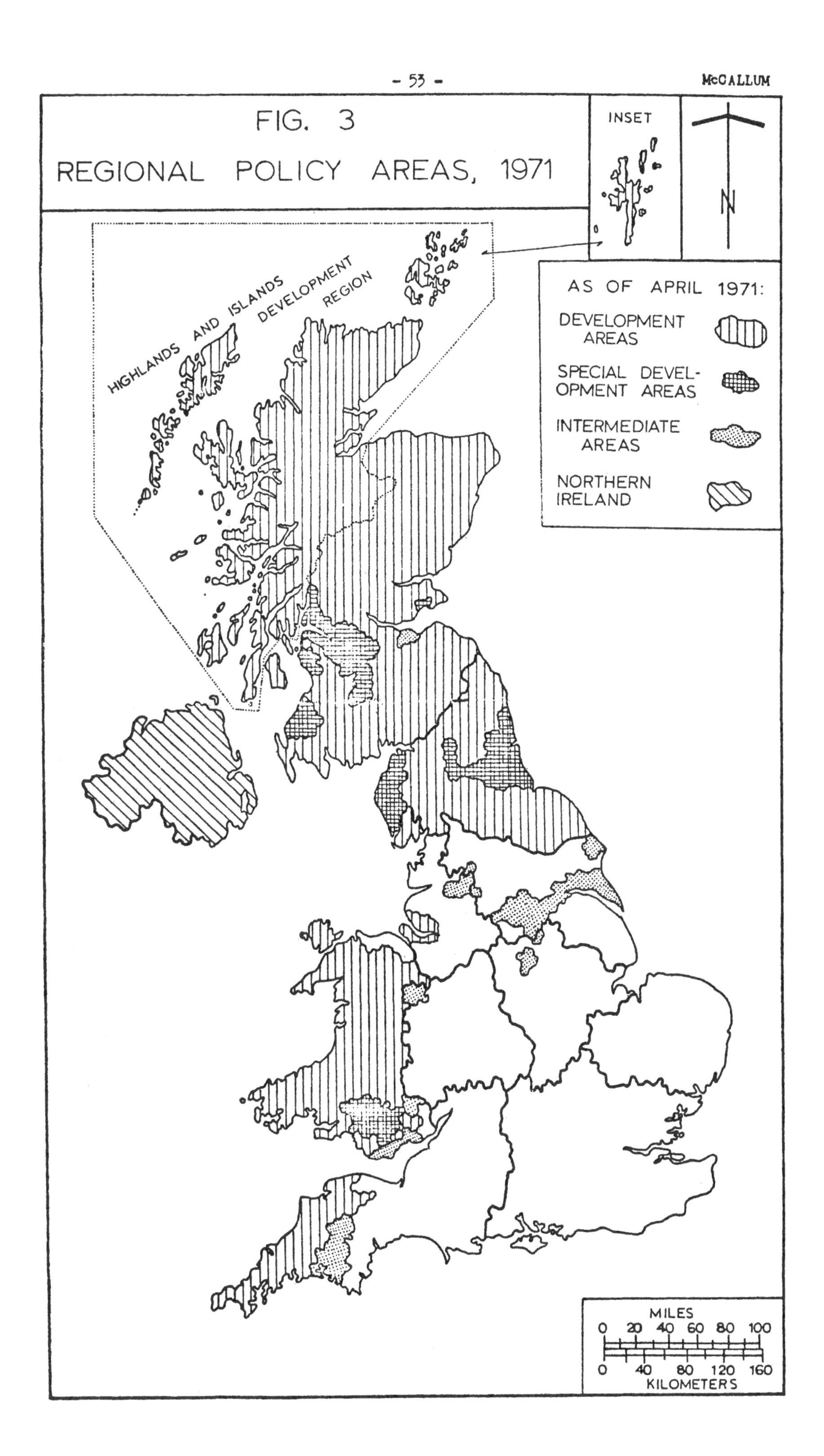
FIG. 3
REGIONAL POLICY AREAS, 1971
INSET
N
HIGHLANDS AND ISLANDS DEVELOPMENT REGION
AS OF APRIL 1971:
DEVELOPMENT AREAS
SPECIAL DEVELOPMENT AREAS
INTERMEDIATE AREAS
NORTHERN IRELAND
MILES
0 20 40 60 80 100
0 40 80 120 160
KILOMETERS

FOOTNOTES

1. Certain aspects of British political structure should be kept in mind while examining national regional policy. The United Kingdom (U.K.) is a unitary state governed by the national Parliament in London; its principal constituents are England, Wales, Scotland, and Northern Ireland. The terms "Britain" or "British" refer loosely to the U.K. or parts thereof; but the term "Great Britain" (G.B.) refers specifically to England, Wales, and Scotland only. Northern Ireland has a separate parliament (Stormont) with broad responsibilities for local affairs, including most of regional policy. Because of Northern Ireland's unique independence in this field, discussions of national regional policy usually concern Great Britain only. Within Great Britain itself, Scotland has a special constitutional position which has secured her a separate administrative structure (the Scottish Office) headed by a Cabinet Minister. But while Scottish affairs are usually discussed and voted on separately in Parliament, national regional policy tends to be the same north and south of the border. Similarly, although Wales is treated separately on some matters -- and has recently acquired a separate Welsh Office -- on matters of regional policy it is ordinarily not treated differently. Within England itself there are no constitutional or practical divisions of national consequence, though the future may eventually bring some devolution of power.

2. Expenditures on British regional policy, calculated as funds expended in the Development Areas deliberately in excess of those expended elsewhere under the same programs (and including extra Public Works Loans), totalled around £181 million ($434 million) in 1967/68 and £305 million ($732 million) in 1968/69. (See D.E.A., 1969b, p 97 and E.F.T.A., 1970, p 48) These are quite large sums in absolute terms, especially for a nation with only about one-quarter of the population and about two-fifths the per-capita G.N.P. as the United States. For an expenditure in the U.S. to achieve the same relative scale, moreover, it would have to total around $7,000,000,000 for 1968/69 to equal the British policy's 0.8 per cent share of G.N.P.

3. The theme of resistence to migration is one that continues throughout the history of British policy. People have often been loath to migrate long distances, even in time of acute distress. Ministry of Labour investigators in 1934 noted this repeatedly, emphasizing the strong attachment of the worker to his locale and even citing instances of workers transferred from the North of England who left their new jobs in Kent (the far South) and walked back to their homes (Ministry of Labour, 1934). Even today, with society less tradition-bound and yet with inter-regional differences in job opportunities still significant, the gross amount of inter-regional migration is only equal to about 1½ per cent of the population in any given year -- and this in a country whose inter-regional distances are trivial by U.S. standards (Brown, 1969, p 774). McCrone (1969) discusses this point at some length, and Odber (1965), emphasizing its importance as a determinant of public reaction to regional policy measures, judges it to be an important complication in later policy deliberations. Economists writing on the regional "problem" have been particularly intolerant of this persistent human trait, as if angered by reality's failure to fit their _a priori_ models of perfect mobility.

4. In March 1933, for example, the national average unemployment rate was 61 per cent in shipbuilding, 35 per cent in metal manufacture, 26 per cent in coal mining, and 24 per cent in textiles (McCrone, 1969, p 92).

5. The 1929 De-Rating Act -- intended to relieve the burden of local authority property taxes in hard-hit areas through central government assistance -- might be viewed as an earlier regional policy measure. But while this was designed as assistance for depressed areas, it was not at the time conceived of as a "regional" policy but rather as a financial one. In any event, it was a fairly weak measure of little effect.

6. As strange as it may appear to the modern eye, the 1934 investigators laid unanimous stress on the movement of unemployed urban workers to small-holdings on the land as a means of alleviating distress. While this proved of little effect on the total problem, it was often warmly endorsed by the workers themselves. In South Wales there is still a tradition of miners maintaining small garden-farms as supplements to their wage incomes.

7. Though there was apparently no adverse reaction to this step, tax incentives were dropped from the industrial incentive powers in the post-war regional legislation, not to reappear until 1963.

8. The successful large industrial estates of this early period of public support were Team Valley (Northern England), Treforest (Wales), and Hillington (Scotland). See European Free Trade Association, 1970.

9. The "Public Health" and "Garden City" traditions of early town planning had contributed to the widespread acceptance of this idea and of the related idea of new towns. It is a tribute to the persuasiveness of these early reformers that Barlow assumed the evils of large, crowded cities to be "so generally recognized as to need no comment".

10. This dichotomy between 'urban' and 'regional' characterized British planning in the early post-War years. Measures for urban redevelopment, including all the paraphernalia of new towns, were conceived of in terms of local decentralization and provision of housing in a "good" environment; institutionally, these programs and basic outlooks became entrenched in the Ministry of Housing and Local Government (and its variously-named predecessors and successors). Measures for regional development, on the other hand, focused on industry and manipulation of private location decisions; they were institutionally based in the Board of Trade. The divergence of view between these two central government agencies was sometimes highly pronounced, as in new towns policy: each looked to new towns to serve a different purpose and so they often worked against one another. This split in responsibility (as well as in outlook) has never been fully resolved, even though the 1970 reorganization of Ministries did bring most of the two together under the over-all control of the Department of the Environment. During most of the Fifties and Sixties the conflicts between these two Ministries were well known and widely commented upon. See also footnote 13 and the text discussion about changes in approaches to new towns about the time of the 1963 White Papers.

11. Odber, Allen and Bowden (1957) observed that "The White Paper was, however, careful not to spoil the simple charm of the phrase "proper industrial balance" by trying to define it...". The idea of "balance" became (and remains) a fixture in regional discussions, whether talking about a balanced industrial mix within a region or about the balance among regions in employment and population. But the concept remains about as loosely defined as in 1944.

12. Additional Development Areas were designated later under this same Act: Wrexham and Wigan/St Helens in 1946; Merseyside, the Highlands and Islands of Scotland, and the North East of Lancashire in 1948. In the case of the Highlands and Islands, however, only a small area was designated, the idea being to concentrate development instead of dispersing it over the vast spaces of Northern Scotland; this is an interesting early use of the growth center idea in regional policy.

13. The new towns in fact proved capable of offering a powerful counter-attraction to the Development Areas because the accepted orthodoxy of the new town idea strongly insisted on 'balance' (as many jobs as resident workers) to avoid the creation of "bed-room suburbs". In line with the new towns' priority position as suppliers of much-needed housing there was accordingly an unwillingness on the part of authorities to restrict the I.D.C.s in the new towns. While an industrialist might not get permission to build in London, then, he could probably get both permission and a planned industrial estate — all within a few miles of London — simply by going to the new towns. In many cases this resulted in a decentralization from London yet no net gain for the other regions; in effect, the new towns were thus set up in competition with the Development Areas. See Thomas, 1969, Chapter 9 and also Rodwin, 1970, p 123. An exception was Scotland, where a number of new towns quickly became important centers of attraction for new industry.

14. Buildings over 5,000 square feet only (smaller buildings did not have to be approved. Figures from Odber, Allen and Bowden, 1957, Table V, p 17. It must be remembered that "approvals" did not always lead to actual construction; and when it did there could still be a time lag of several years. Thus the statistics for "approved area" and new area actually built will be quite different.

15. Calculated at $2.80 per pound sterling, the exchange rate which held from the autumn of 1949 to the autumn of 1967.

16. It is likely, however, that select parts of the civil service managed throughout the early Fifties to keep alive some of the ideas behind much of the earlier work, despite the public statements of its political masters. The British civil service has tended to function like this to preserve the "continuity" of government despite changes in whatever direction. In addition, there is the simple fact that the original issue itself (regional imbalance) remained alive, even if studiously ignored.

17. Advisory regional plans were commissioned and prepared in the mid-Forties for both Greater London and the Clyde Valley (Greater Glasgow) by Abercrombie (1945, 1949). Though principally physical plans with a town planning orientation, they were important foundation efforts which might have been of much more substantial long-term benefit had they been extended,

supplemented, and implemented as part of a coherent planning operation. Even so, the Clyde Valley plan provided for many years a basis for location decisions such as new towns, major highways, principal housing developments, etc. Limited as such a 'physical' plan might have been, it was nonetheless an important pioneering work in planning of potentially great value if used in conjunction with other aspects of comprehensive regional development.

18. Calculated for both periods at $2.80 per pound sterling.

19. Some of these industries suffered a basic secular decline (coal-mining, ship-building, railroads) while others suffered more from cyclical distress (metal manufacture, heavy engineering). But both types suffered throughout the later Fifties whenever conditions called for deflation or economic restraint.

20. Taken from Odber (1965), Table 4, page 401. These are old (pre-1964) Standard Regions and not the new (post-1964) official Economic Planning Regions, though in many cases the areas are the same. British unemployment is calculated on a quite different basis from that used in the U.S. and reflects only those registered on official rolls; this makes it perhaps 1/3 to 1/2 lower than the U.S. figures, which are based on sample survey. This must be kept in mind when comparing figures from the two countries.

21. If timed correctly, the recovery could be coordinated with the calling of a General Election, allowing the Government to compaign on a basis of rising prosperity. The Conservatives successfully operated this manoeuver in 1959 (Bailey, 1968).

22. Odber's discussion (1965) of the interest rates charged by the Board of Trade Advisory Committee (BOTAC) on loans to enterprises in Development Areas is a useful illustration of the Treasury attitude. BOTAC was anxious to keep rates low, in order to stimulate investment; and they naturally disliked having to raise their rates every time the Treasury did, particularly in periods of recession. "The gave the impression of wanting to keep their recommended rates down a little when bank rate rose, but of being acutely conscious that if they stepped outside "pretty narrow limits" they would be overruled by the Treasury. This point was confirmed by a Treasury official." Odber judges this to be typical of the Treasury attitude: "Thus the Treasury mind translates a requirement to charge no more than a commercial rate into a doctrine of charging no less than a commercial rate."

23. In general, the "Stop-Go" type of policy aggravates the problems of the poorer regions. In a "Stop" they are usually hit harder, being more vulnerable. And in the "Go" they are usually slower to recover, having less resilience and farther to go. Sometimes they never even recovered fully before being hit with the next "Stop".

24. McCrone (1969) relates the 1958 Steel Mill decision as an example of the piece-meal and poorly thought-out kinds of solutions the Government was attempting at this time. "The steel industry which had been planning a new strip mill project, originally to be sited in the Midlands, found its plans subject to much canvassing from hard hit regions. At length, as a result of personal intervention on the part of the Prime Minister in 1958, the project was split, one strip mill being set up at Llanwern in South Wales and another at Ravenscraig in the Central Belt of Scotland. There is little sign that the economic implications of this were seriously considered...

Experience has shown that in consequence neither project was large enough to enjoy the economies of scale which the large integrated works on the Continent now enjoy, and in the case of the Scottish mill especially, which was situated inland, the siting is now poor." (p 118) This last item is particularly significant, for the Clyde estuary is the best deep-water available in Great Britain and a site at Hunterston is being suggested in 1971 for a brand new ore terminal. Such a new terminal would be capable of berthing the deepest-draft boats afloat or envisioned -- yet the ore will still have to be trans-shipped overland to the Ravenscraig mill. The kind of muddle represented by the steel mill decision stemmed from the Government's lack of any over-all strategy or plan to guide regional policy decisions. The Government had accepted the political need to act; but it had not accepted the need to act intelligently in line with some well thought-out policy.

25. It must be remembered that despite the ups and downs of national regional policy thinking in London, Scotland's administrative independence has allowed her considerable freedom to pursue a different course. During the 1950s, when regional planning was moribund nationally, there still existed a nucleus of regional concern within Scotland. This "continuing presence of a team of planners concerned with the problems of the region as a whole helped to make Scotland the leader in many aspects of regional planning" (Harris, 1966b, p 254). The Scottish new town of Cumbernauld, for example, was the only one to appear in Britain during this period. Harris also credits this Scottish activity with, among other things, the birth of the growth area idea in Britain. The 1961 "Toothill Report" (Scottish Council (Development and Industry), 1961) is yet another instance of Scottish initiative; it "played a considerable role in refocussing national views on what needed to be done. It influenced the two White Papers prepared in 1963 dealing with Central Scotland and the North-East, and a series of subsequent regional studies for the South East, Wales, the North West, and the West Midlands." (Rodwin, 1970, pp 138-139). Conditions Favourable to Faster Growth, for instance, frequently cited the Toothill Report and emphasized its role as the first such exercise in the country. In some ways this leadership still exists, with Scottish regional institutions being unusually strong and active in comparison with those which have been set up elsewhere.

26. The year 1963 also brought Lord Hailsham's much publicized "fact-finding" visit" to the depressed North East of England, during which tour he donned a cloth cap (the distinctive apparel of the British working-man) to demonstrate to the press photographers how concerned he and the Government were about unemployment there.

27. At the 1959 General Election the political implications had been clearly spelled out. In the nation as a whole the Tories increased their majority by 21 seats; from 54.6% of seats in 1955 they went up to 58.0% in 1959. In Scotland, however, the Tories lost 5 seats, dropping from 50.7% of seats (1955) down to only 43.7%. Similarly, in the North of England the Government in 1959 gained only 2 seats over the 1955 total, the smallest percentage gain in any English region.

28. The break-through here was the establishment in 1963 of the National Economic Development Council (N.E.D.C.). While not an organ of central planning in a control sense, it was designed to assist the Government in

relating its economic policies to one another and to reasonable expectations of future developments in the private and public sectors. See: Bailey, 1968; Hagen & White, 1966; Hutchison, 1968; Mitchell, 1966; and Shone, 1966.

29. Loasby (1965) argues that the 'growth area' idea -- which figures importantly in both White Papers -- was first suggested for Scotland in 1952 in the Scottish Council's Report of the Committee on Local Government in Scotland and first suggested for the North East in 1957 by Odber, Allen and Bowden.

30. The prevailing idea was typically expressed by Mr. Douglas Jay (later President of the Board of Trade) in March 1963: "Town and country planning and economic planning are quite distinct. Town and country planning is concerned with the use of land." (Harris, 1966a, p 93). This attitude, moreover, was not the result of ignorance or indifference. Quite the reverse, it was (and still is) strongly supported by the entrenched professional body of British planners, the Town Planning Institute, which has scarcely begun to acknowledge that planning involves anything more than architecture, engineering, and law.

31. At the 1964 General Election the Conservatives won 48.3% of all seats in Parliament. However, in Scotland, Wales, and Northern England (here including most of the Northern, North Western, and Yorkshire and Humberside Planning Regions) the Conservatives won only 30.3% of the seats, in sharp contrast to the 62.1 per cent they won in the rest of the country. In terms of the change from the last General Election, the Conservatives in 1964 lost 27.8% of their 1959 seats in the Scottish, Welsh, and Northern English regions, while they only lost 11.6% in the rest of the country. If the Tory drop had been this same 11.6% over the whole country, they would have stayed in power in 1964. In this sense it was the depressed areas which turned them out.

32. The idea of a national Ministry of Economic Affairs was first brought up during the post-war Labour Government, when an effort was made in 1947 to set up such a ministry under Sir Stafford Cripps; but no official establishment was made, and Cripps moved on to become Chancellor of the Exchequer. Another brief attempt was made in 1950 under Gaitskell, but this too resulted in nothing permanent and ended with Gaitskell's move to Chancellor. The idea of such a ministry has always been a strong article of faith in a Labour Party suspicous of City influence in the Treasury, and so it was no surprise that the D.E.A. was quickly set up in 1964. It is ironical, though, that the D.E.A. was abolished, in October 1969, by the very same Labour Government which had set it up five years earlier. See Bailey, 1968; Hagen & White, 1966; Mitchell, 1966; and Roll, 1966.

33. As might be expected of any attempt at an all-purpose regional definition, there arose many difficulties which could only be resolved by an arbitrary decision admittedly leaving some anomolies in the system. Nonetheless, the division seems to have worked as well as any single regionalization can be expected to. Incidentally, this might remind some U.S. readers of the attempts at systematic regionalization and establishment of uniform planning and administrative areas made by the National Resource Committee in the mid-Thirties.

34. Councils of "leaders" and boards of government officials are sometimes used in the U.S. in areas which lack a formally constituted government structure and which require cooperative effort. Many ad hoc activities -- notably metropolitan transportation studies -- have used such a set-up.

35. During the thirteen years of Conservative Government 1951-1964, a total of 49 advance factories were authorized for the Development Areas -- less than 4 per year. In the 5½ years from October 1964 to April 1970 the Labour Government authorized 221 -- roughly 40 per year. (Labour Party, 1970, p 20)

36. Both Governments had attempted to lead the way by decentralizing some government offices out of London. In most cases these moves went to the Development Areas, thereby both reducing the concentration in London and providing welcome office employment in the depressed areas. In the early and mid Sixties, the Post Office Savings Bank went to Durham (Northern Development Area), the National Savings Office to Glasgow (Scottish Development Area), a new mint to South Wales (Welsh Development Area), and GIRO to Bootle (Merseyside Development Area). It is clear, however, that government employment is still massively concentrated in London and the South East and that the enormous potential of such decentralization has scarcely been touched so far. The Location of Offices Bureau in the years between 1965 and 1971 had only limited accomplishments; the vast majority of offices it managed to get out of London went only to suburban locations in the outer metropolitan area or, at best, to towns within the South East. Like the new towns policy, location of office policy has restricted London only to the benefit of areas near London

37. The strong deflationary measures taken in July 1966 -- and added to almost annually since -- had the expected effect of immediately slowing growth and raising unemployment. Unfortunately, this did not bring a balance of payments recovery, and devaluation became necessary in November 1967. But even with the temporary boost of devaluation matters have not improved enough to allow for a reflation, at least in the view of the Treasury and the ruling economic powers; in 1970 and 1971, though, inflation has replaced the balance of payments as the excuse for keeping on the "brakes". Accordingly, the economy has been in a deliberately restrained, defated state since mid-1966 continuously, and unemployment has been unrelievedly high and getting higher. This state of the economy must be kept in mind when considering regional policy measures in the late Sixties, for all during that time the Government has been attempting to secure regional growth and regeneration in a context of national stagnation.

38. The seven counties of the Highlands and Islands share, among other things, the historical tradition of crofting -- a peculiar form of small-holding agriculture -- and (Orkney and Shetland excepted) a common background of Gaelic culture and social organization. Moreover, all the counties suffer the disabilities of geographical isolation, transportation difficulties, and the lack of an urban, industrial base. The total 1968 population was only about 275,000, representing about 0.5% of the U.K. total; but the area comprises almost 15% of the total area of the nation. The population density of the Highlands and Islands is about 7.6 persons per square kilometer as opposed to 265 per square kilometer in the rest of the U.K. Economically, the region is one of the most depressed in the nation. Income per head was only 66% of the U.K. average in 1964/65 and only 80% of the over-all Scottish average. Unemployment is higher, ranging from

6.9 to 8.1 per cent in 1966-68 as compared to 2.9 to 3.9 in Scotland as a whole and 1.5 to 2.5 in Great Britain. In addition, a high rate of migration (4.8 per 1000, 1961-68) has been combined with a low rate of natural increase (3.6 per 1000) to yield a steadily declining total population and an unbalanced age distribution. The region is thus an underpopulated and declining rural area within a very densely populated and growing industrial nation. It is probably only the intractable nature of Highland problems, and the patent failure of numerous past efforts, that moved the Government to establish such a relatively novel body as the Highlands and Islands Development Board.

39. After surviving a considerable storm in its early years the Board has generally consolidated a position of strength within the Highlands. The change of Government in June 1970, however, caused speculation about the future of the Board under a Conservative Government not conspicuously friendly to such manifestations of governmental activity.

40. "Overspill" schemes are formal arrangements "made under the Town Development Act 1952, and the Housing and Town Development (Scotland) Act 1957, between the local authorities of large industrial cities without enough land at their disposal and those of 'expanding' towns, for planned voluntary transfers of population and industry. London, Birmingham, Glasgow, Liverpool and Manchester are examples of cities which are 'exporting' under overspill agreements" (Central Office of Information, 1968, p 23). The need to 'export' is usually the result of physical redevelopment in densely populated urban areas whereby the population displaced is much greater than that which can be rehoused at acceptable densities.

41. While the regional differential was thus ostensibly 20%, it in practice worked out rather lower, due to two factors: first, the delay in payment, which tended to be at least a year; and second, the ruling by which tax allowances were permitted only on the non-grant portion of the capital expenditure. This can be illustrated by a hypothetical example (courtesy of Professor Thomas Wilson):

	Development Areas:	Rest of Country:
Investment Grant:	40.0%	20.0%
1 year time lag @ 10% reduces grant to:	36.4%	18.2%
Calculate tax allowance on remaining 60% and 80% respectively:	16.6%	22.1%
Add, to get Total Benefit:	53.0%	40.3%

Thus the difference in total benefit is only 12.7%, and not the apparent 20.0%. It is important to remember such discrepencies between real and apparent value of incentives when evaluating instruments of policy. It is also of interest that Northern Ireland paid capital grants immediately and thus provided a rather better incentive.

42. These five -- the Scottish, Northern, Merseyside, Welsh, and South-Western Development Areas -- remain the basic geographical units for regional policy, with the 1970 addition of the Intermediate Areas.

43. The tax concession method has another draw-back: it is biased in favor of large multi-plant firms in which losses at any one branch plant can be merged with the whole company's profits to produce a situation in which the firm can still benefit from tax concessions. This 'trick' is not open to small one-plant firms and thus the system works against the new or small entrepreneur.

44. Though a commonly accepted view, this thesis was challenged in a recent article by Chisholm (1970) in which he judges the case "not proven either way". There did seem a tendency for large unit industries (such as petroleum and chemicals) which were expanding generally throughout the period to go to the Development Areas and take advantage of the grants. But it would be unwise to generalize from the basis of these few, if conspicuous, examples. Related to this is the controversy over British regional policy and GATT (General Agreement on Trades and Tariffs). The large sums of money which regional policy makes available, especially to capital intensive industries, may be viewed as a way of evading GATT restrictions on unfairly subsidized competition. The Norwegians in particular have made this objection with respect to the large-scale aluminum plants being build (with sizeable grants) in Wales and Scotland.

45. This could be an important point, for some argue that 'efficiency wages' are in fact higher in the depressed regions, at least in some industries, because of this artificially boosted level of money wages. REP, then, could provide a way of equalizing 'efficiency wages'. On a slightly different point, REP can also be attacked as an evasion of GATT restrictions. With virtually all of Britain's ship-building industry being in Development Areas, the REP subsidy could easily be seen from abroad as a surreptious subsidy to Britain's ship exports.

46. It is reasonable to expect a Labour Government to be responsive to such distress, as no other areas in Britain have been (or remain) so solidly Labour as the traditional coal-mining regions; in addition, the problems of the mines have always occupied a special place in the history and ideology of the Labour Party.

47. It is not unfair to suggest, though, that the Labour Government had already quietly dropped the 'growth center' and concentration ideas in favor of a much more broad-scale approach. Such a move was probably motivated by the worsening economic situation and the political or welfare need to secure as much assistance for as much of the depressed areas as possible

48. Related to this are other efforts to accomodate the drastic decline in coal-mining. County Durham has attempted, since its Development Plan of 1951, to rationalize the settlement pattern by deliberately running down and (ultimately) eliminating many of the small, uneconomical old mining villages which could no longer be supported by colliery employment.

49. The fate of the D.E.A. was tied up in a number of complex issues. For one thing, it became associated with internal struggles within the Labour Party for leadership, and with Prime Minister Wilson's desire to bring more of the control of economic policy under his personal command. It was also tied up with efforts to free the country (as Labour saw it) from the immovably conservative economic control of the Treasury; the D.E.A. had been one of their moves to reduce the Treasury's independent position. But as a result of Labour Party personalities and politics, and of economic and financial crises, the D.E.A. never really had a chance to establish itself, its powers gradually whittled away and circumscribed until, in 1969, it was formally ended.

50. That this can be a powerful tool is suggested by the location of the large new aluminum works at Invergordon in Northern Scotland; the as yet undisclosed special terms for large-volume electric power supply were probably a major factor in the location decision.

51. All of the measures available in the Development Areas are also available in the Special Development Areas, though supplemented by certain special extra aids.

52. This figure includes all regionall-differentiated programs, along with the preferential access to the Public Works Loan Board. See footnote 2, above.

53. At the end of the election the Conservatives held 330 seats, Labour 288, Liberals 6, and Others 5. There were marked regional variations in the results. The Development Areas remained solidly Labour (107 seats out of 156), and the Conservatives gained only 9 seats over their 1966 total. Similarly, in the Intermediate Areas the Tory gain was only 3 seats and Labour remained firmly in the majority (25 seats out of 32). In the remainder of the U.K., however, the Tories jumped into a commanding lead (281 seats to 156 for Labour), gaining 64 seats. Compared to 1966, Labour in 1970 lost only 7% of its seats in Development and Intermediate Areas; but in the remainder of the country they lost 29%. In this sense it was the prosperous areas which turned Labour out.

54. The Board is an innovation strongly opposed in certain sections of the Conservative Party. It is especially disliked by old-style "lairds" and "Backwoodsmen", though the Government has vaguely committed itself to continuing support for the Board. But it is safe to suggest that, had the Highland Board not strongly established itself in its first five years, the change of Government would have spelled its end just as surely as it did the Northern Pennines Rural Development Board's. See below.

55. No apparent reasons for this closure can be given other than the Conservatives' instinctive dislike of such 'irregular' governmental activities -- especially when they impinge upon land-owning. But having cut off the Board after only a year and a half they have excluded the possibility of finding out whether or not such an approach can be useful or effective. One particularly unfortunate by-product of this decision is the potential loss of the rural bus services in the area; the Rural Development Board had taken upon itself to subsidize the rural routes. Without that (or some similar) assistance, these routes -- which usually provide the only form of transport available for the scattered population -- will be withdrawn, as they are inescapably non-commercial.

56. The effect of this sudden change is difficult to assess, but some large-scale projects then in the negotiating stages were halted; a few may eventually be dropped altogether. In any case, a time of low profits, low corporate liquidity, and high interest rates is a disadvantageous time to switch incentives, since tax allowances are much less effective in such circumstances.

57. It was suggested that the limitation will affect "only the more capital intensive types of project", implying that except for conspicuous cases (like petro-chemicals) grants will not be materially affected by employment creation criteria — the reverse of the Conservatives' pre-election stance.

58. Data are for public investment in new construction in roads, fuel and power, hospitals, dwellings, education, and other social and environmental services. Per capita figures are calculated from Table 2, Regional Planning Policy (Labour Party, 1970).

59. There are reports that during its last year in office Labour had decided, in consultation with the Treasury, to "cool off" the high level of differential infrastructure spending, especially in areas like Scotland which have enjoyed a particularly favored position. Insofar as the new Government is likely to go along with such a plan, they may end up spending even less than the previous Government.

60. The current economic crisis is not, for once in the post-war history of Britain, a balance of payments problem; in fact, the country has enjoyed a positive balance throughout 1969 and 1970. Equally, inflation — though an important complication — is patently not the central problem. Britain is suffering from what looks very much like an orthodox recession, with high and rising unemployment, high rates of corporate failure, low rates of investment, stagnant or declining productivity, and slow or zero over-all growth.

61. Figures are for wholly unemployed, excluding school-leavers, annual averages of monthly percentages. These are the new (post-1964) official Economic Planning Regions; before 1965 there was a different set of regions in parts of the country. Data are from Abstracts of Regional Statistics, Monthly Digest of Statistics, and Department of Employment Gazette.

62. The announcement of West Central Scotland's designation was rushed through two weeks earlier than planned so that it could come just before the public announcement of the Rolls-Royce collapse. Rolls-Royce was at that time Scotland's largest private employer and the Government obviously felt that they had to somehow cushion the shock of its demise.

63. In the Parliamentary discussion on 18 February 1971 following the Government's official announcement of the Special Development Area designations, Labour M.P.s argued that the amount of assistance was merely being raised back up to what it had been under their own Government, while the Conservative speakers insisted that additional expenditure was involved. However, because so many of the items are estimated and difficult to measure, the issues remains unclear.

64. These regional policy initiatives, combined with the "nationalization" of Rolls-Royce, caused considerable unrest in the right wing of the Conservative party. Unavoidable as these steps may have been in the circumstances, they still ran rough-shod over certain party ideologies.

65. At this time Mr Davies also announced that the Government would not be publishing a White Paper on regional policy; he said that the matter was already quite clear. (Guardian, 9 March 1971) The Government had on several earlier occasions -- including in the October White Paper on "Investment Incentives" -- promised a "thorough-going study of regional development policy"; but no such study or White Paper ever made an appearance. In a Sunday Times article in February 1971 it was reported that the long-lost White Paper had been surpressed due to intense criticism within the civil service where it was being circulated. The implication was that it had been far too ideological in cast and had, in any event, been over-taken by reality. However, parts of such a study had probably been available within the Government when it was formulating its policies in January and February, even though nothing official was said publicly. Finally, it might be thought that the sweeping nation-wide victories of the Labour Party in the May local government elections, and their victories in by-elections, had cautioned the Government about any further negative changes.

66. It must be admitted, of course, that administrative reorganizations do not necessarily end or even reduce inter-departmental conflicts; they may simply internalize them without substantially affecting any of the personalities, attitudes, or established 'empires'.

67. Two of the districts, Orkney and Shetland, will in many respects be treated as separate regions, in view of their considerable isolation. Also, in three of the sparsely populated regions the districts will not exercise full powers, some of their functions (local planning, building control) being given up to the region. The system is thus being flexibly applied, as any system must be in an area as varied as Scotland.

68. The creation of these metropolitan governments is an important innovation, carried farther in the White Paper (6 metropolitan counties) than in the Royal Commission Report (which recommended only 3). If the experience of the Greater London Council set-up is a guide, the new governments should prove adequate and perhaps quite successful. And with the Government's plan all of the English conurbations and large cities would be inside some metropolitan organization -- only one city over 400,000 (Bristol) and four over 250,000 (Nottingham, Hull, Leicester, and Stoke) would remain outside.

69. In the process, however, 23 large county boroughs over 100,000 population will lose their independent status, as well as the full range of local government functions which they have been and are now exercising; this is a feature which makes these cities extremely apprehensive of their future fate under an often rural-dominated county government.

70. There has always been some degree of split in British local government between Labour city councils and Conservative county councils; in the Midlands and South of England particularly, the "county interests" have always been very powerful within the Conservative party. It was widely feared that a Tory Government would prove susceptible to pressure from those quarters; so it was no surprise when the new proposals followed very closely

the suggestions of the counties while largely ignoring those of the cities, of the Redcliffe-Maud Commission, and of the Commission minority report. What has emerged is politically expedient, with some good features, but far from a fundamental change in most respects.

71. Figures are estimated, partially on the basis of McCrone's (1969) and E.F.T.A.'s (1970) data, supplemented by subsequent individual program announcements. For each year, the figures would be, roughly: £156 million for 1967/68, £265 million for 1968/69, £304 million in 1969/70, and £325 million in 1970/71. All such estimates, however, are problematical and should be taken with caution.

72. British academics must share in the blame. Regional policy oriented investigations only became fashionable in the Sixties and there are many difficult problem areas within which policy-makers would be happy to have more (and more reliable) guidance from the "experts" — especially from the economists.

* * * * * *

NOTE: The author would like to apologize to the reader for the awkwardness and inconvenience of the placement of footnotes at the end of the text. As these footnotes are extensive and, in the author's opinion, useful comments and elaborations upon the main themes of the text, it was hoped that these would be included at the bottom of the relevant pages, where they could be consulted with ease and without interrupting the flow of the article. But the organizers of the conference imposed this present unsatisfactory system upon the contributors.

REFERENCES

ABERCROMBIE, Sir Patrick and MATTHEW, Robert H. (1949). The Clyde Valley Regional Plan 1946. Report prepared for the Clyde Valley Regional Planning Committee. Edinburgh: H.M.S.O., 1949.

ABERCROMBIE, Sir Patrick (1945). Greater London Plan 1944. Report prepared for the Standing Conference on London Regional Planning. London: H.M.S.O., 1945.

BAILEY, Richard (1968). Managing the British Economy. London: Hutchinson, 1968.

BOARD OF TRADE (1963). "The North East: A Programme for Regional Development and Growth". Cmnd 2206. London: H.M.S.O., 1963.

BOARD OF TRADE (1948). "Distribution of Industry". Cmnd 7540. London: H.M.S.O., 1948.

BROWN, A. J. (1969). "Surveys of Applied Economics: Regional Economics, with Special Reference to the United Kingdom". The Economic Journal, Volume LXXIX, Number 316, December 1969.

BROWN, A. J., et al (1968). "Regional Problems and Regional Policy". National Institute Economic Review, Number 46, November 1968.

BROWN, A. J., LIND, H., and BOWERS, J. (1967). "The 'Green Paper' on the Development Areas". National Institute Economic Review, Number 40, May 1967.

CAIRNCROSS, A. K., Editor (1954). The Scottish Economy. University of Glasgow Social and Economic Studies Number 2. Cambridge: Cambridge University Press, 1954.

CAMERON, Gordon C. (1970). "Growth Areas, Growth Centres and Regional Conversion". Scottish Journal of Political Economy, Volume XVII, Number 1, February 1970.

CAMERON, Gordon C. and REID, Graham L. (1966). Scottish Economic Planning and the Attraction of Industry. University of Glasgow Social and Economic Studies Occasional Papers Number 6. Edinburgh: Oliver and Boyd, 1966.

CAMERON, Gordon C. and CLARK, B. D. (1966). Industrial Movement and the Regional Problem. University of Glasgow Social and Economic Studies Occasional Papers Number 5. Edinburgh: Oliver and Boyd, 1966.

CARTER, Charles F. (1969). "The Hunt Report". Scottish Journal of Political Economy, Volume XVI, Number 3, November 1969.

CENTRAL OFFICE OF INFORMATION (1968). Regional Development in Britain. Reference Pamphlet Number 80. London: H.M.S.O., 1968.

CHINITZ, Benjamin (1965). "Regional Economic Policy in Great Britain". Urban Affairs, Volume 1, Number 2, 1965.

CHISHOLM, Michael (1970). "On the Making of a Myth. How Capital Intensive is Industry Investing in the Development Areas?" Urban Studies, Volume 7, Number 3, October 1970.

CLARKE, Sir Richard (1966). "The Machinery for Economic Planning: The Public Sector". Reprinted from Public Administration, Spring 1966.

CULLINGWORTH, J. B. (1967). Town and Country Planning in England and Wales. Second Edition. London: George Allen & Unwin, 1967.

DEAN, Sir Maurice (1966). "The Machinery for Economic Planning: The Ministry of Technology". Reprinted from Public Administration, Spring 1966.

DENNISON, S. R. (1939). Location of Industry and Depressed Areas. Oxford: Oxford University Press, 1939.

DEPARTMENT OF ECONOMIC AFFAIRS (1969a). "The Intermediate Areas: Report of a Committee of Inquiry under the Chairmanship of Sir Joseph Hunt" (The "Hunt Report"). Cmnd 3998. London: H.M.S.O., 1969.

DEPARTMENT OF ECONOMIC AFFAIRS (1969b). "The Task Ahead: Economic Assessment to 1972". London: H.M.S.O., 1969.

DEPARTMENT OF ECONOMIC AFFAIRS (1968). "Economic Planning in the Regions" (Second Edition). London: H.M.S.O., 1968.

DEPARTMENT OF ECONOMIC AFFAIRS (1967). "The Development Areas: A Proposal for a Regional Employment Premium". London: H.M.S.O., 1967.

DEPARTMENT OF ECONOMIC AFFAIRS (1966a). "The Problems of Merseyside: An Appendix to the North West Study". London: H.M.S.O., 1966.

DEPARTMENT OF ECONOMIC AFFAIRS (1966b). "Investment Incentives". Cmnd 2874. London: H.M.S.O., 1966.

DEPARTMENT OF ECONOMIC AFFAIRS (1965a). "The National Plan" Cmnd 2764. London: H.M.S.O., 1965.

DEPARTMENT OF ECONOMIC AFFAIRS (1965b). "The North West: A Regional Study" London: H.M.S.O., 1965.

DEPARTMENT OF THE ENVIRONMENT (1971). "Local Government in England: Government Proposals for Reorganisation" Cmnd 4584. London: H.M.S.O., February 1971.

DEPARTMENT OF TRADE AND INDUSTRY (1970). "Investment Incentives" Cmnd 4516. London: H.M.S.O., October 1970.

DUTT, A. K. (1970). "Regional Planning in England and Wales: A Critical Evaluation: Part II: The London and Birmingham Areas as Case Studies". Plan Canada, Volume 10, Number 3, June 1970.

DUTT, A. K. (1969). "Regional Planning in England and Wales: A Critical Evaluation: Part I". Plan Canada, Volume 10, Number 1, April 1969.

EAST ANGLIA ECONOMIC PLANNING COUNCIL (1968). "East Anglia: A Study". London: H.M.S.O., 1968.

EAST MIDLANDS ECONOMIC PLANNING COUNCIL (1966). "The East Midlands Study". London: H.M.S.O., 1966.

EUROPEAN FREE TRADE ASSOCIATION (1970). Regional Policy in EFTA: Industrial Estates. Geneva: E.F.T.A., March 1970.

EUROPEAN FREE TRADE ASSOCIATION (1968). Regional Policy in EFTA: An Examination of the Growth Centre Idea. University of Glasgow Social and Economic Studies Occasional Papers Number 10. Edinburgh: Oliver and Boyd, 1968.

FOGARTY, M. P. (1945). Prospects of the Industrial Areas of Great Britain. London: Metheun & Co., 1945.

GRIEVE, Sir Robert (1965). "Regional Planning". Journal of the Town Planning Institute, Volume 51, Number 6, 1965.

GRIEVE, Sir Robert (1960). "Regional Planning on Clydeside". Town and Country Planning Summer School 1960. London: Town Planning Institute, 1960.

GRIEVE, Sir Robert (1954). "The Clyde Valley -- a Review". Town and Country Planning Summer School 1954. London: Town Planning Institute, 1954.

HAGEN, Everett E. and WHITE, S. F. T. (1966). Great Britain : Quiet Revolution in Planning. Syracuse, N.Y.: Syracuse University Press, 1966.

HAMMOND, Edwin (1968). An Analysis of Regional Economic and Social Statistics. Durham: Rowntree Research Unit, University of Durham, 1968.

HARRIS, Donald (1966a). "Regional Planning in its Context". Official Architecture and Planning, Volume 29, Number 1, January 1966.

HARRIS, Donald (1966b). "Regional Planning out of the Doldrums". Official Architecture and Planning, Volume 29, Number 2, February 1966.

HARRIS, Donald (1966c). "The Regional Problem: Why Interfere?". Official Architecture and Planning, Volume 29, Number 3, March 1966.

HARRIS, Donald (1966d). "Regional Planning: The Idea of the Growth Area". Official Architecture and Planning, Volume 29, Number 4, April 1966.

HARRIS, Donald (1966e). "The Task of Regional Planning". Official Architecture and Planning, Volume 29, Number 5, May 1966.

HEMMING, M. F. W. (1963). "The Regional Problem". National Institute Economic Review, Number 25, August 1963.

HIGHLANDS AND ISLANDS DEVELOPMENT BOARD (1969). Third Report 1968. Inverness: H.I.D.B., 1969.

HOLMANS, A. E. (1964). "Restriction of Industrial Expansion in South-East England: A Re-Appraisal". Oxford Economic Papers (New Series), Volume 16, Number 2, July 1964.

HOWARD, R. S. (1968). "The Movement of Manufacturing Industry in the United Kingdom 1945-65". London: Board of Trade, 1968.

HUTCHISON, T. W. (1968). Economics and Economic Policy in Britain 1946-66. London: George Allen & Unwin, 1968.

JAMES, John R. (1966). "Regional Planning in Britain". Planning for a Nation of Cities, S. B. Warner, Editor. Cambridge, Mass: M.I.T. Press, 1966.

LABOUR PARTY (1970). "Report of the Study Group on Regional Planning Policy". London: Labour Party, 1970.

LEAN, W. (1967). "National, Regional, and Physical Planning". Journal of the Town Planning Institute, Volume 53, Number 6, 1967.

LEWIS, Ben W. (1952). British Planning and Nationalization. New York: The Twentieth Century Fund, 1952.

LOASBY, B. J. (1967). "Making Regional Policy Work". Lloyds Bank Review, January 1967.

LOASBY, B. J. (1965). "Location of Industry: Thirty Years of 'Planning'". District Bank Review, Number 156, December 1965.

LOCATION OF OFFICES BUREAU (1969). Annual Report 1968-69. London: L.O.B., 1969.

LUTTRELL, W. F. (1964). "Industrial Location and Employment Policy". Town and Country Planning Summer School 1964. London: Town Planning Institute, 1964.

LUTTRELL, W. F. (1962). Factory Location and Industrial Movement. London: The National Institute of Economic and Social Research, 1962.

McCRONE, Gavin (1969). Regional Policy in Britain. University of Glasgow Social and Economic Studies Number 15. London: George Allen & Unwin, 1969.

McCRONE, Gavin (1967). "The Application of Regional Accounting in the United Kingdom". Regional Studies, Volume 1, Number 1, 1967.

McCRONE, Gavin (1965). Scotland's Economic Progress 1951-60. University of Glasgow Social and Economic Studies Number 4. London: George Allen & Unwin, 1965.

McGUINESS, J. H. (1968). "Regional Economic Development -- Progress in Scotland". Journal of the Town Planning Institute, Volume 54, Number 3, 1968.

MEYER, John (1963). "Regional Economics: A Survey" American Economic Review, Volume LIII, Number 1, Part 1, March 1963

MINISTRY OF HOUSING AND LOCAL GOVERNMENT (1964). "The South East Study 1961-1981". London: H.M.S.O., 1964.

MINISTRY OF LABOUR (1934). "Reports of Investigations into the Industrial Conditions in Certain Depressed Areas" Cmnd 4728. London: H.M.S.O., 1934.

MINISTRY OF LABOUR (1935). "First Report of the Commissioner for the Special Areas (England and Wales)" Cmnd 4957. London: H.M.S.O., 1935.

MINISTRY OF LABOUR (1936a). "Second Report of the Commissioner for the Special Areas (England and Wales)". Cmnd 5090. London: H.M.S.O., 1936.

MINISTRY OF LABOUR (1936b). "Third Report of the Commissioner for the Special Areas (England and Wales)". Cmnd 5303. London: H.M.S.O., 1936.

MINISTRY OF LABOUR (1937). "Report of the Commissioner for the Special Areas in England and Wales" Cmnd 5595. London: H.M.S.O., 1937.

MINISTRY OF LABOUR (1938). "Report of the Commissioner for the Special Areas in England and Wales" Cmnd 5896. London: H.M.S.O., 1938.

MINISTRY OF RECONSTRUCTION (1944). "Employment Policy" Cmnd 6527. London: H.M.S.O., 1944.

MINISTRY OF WORKS AND PLANNING (1942). "Report of the Committee on Land Utilisation in Rural Areas" (The "Scott Committee"). Cmnd 6378. London: H.M.S.O., August 1942.

MITCHELL, Joan (1966). Groundwork to Economic Planning. London: Martin Secker and Warburg, 1966.

NATIONAL ECONOMIC DEVELOPMENT COUNCIL (1963a). "Conditions Favourable to Faster Growth". London: H.M.S.O., 1963.

NATIONAL ECONOMIC DEVELOPMENT COUNCIL (1963b). "Growth of the United Kingdom Economy to 1966" London: H.M.S.O., 1963.

NEEDLEMAN, L. and SCOTT, B. (1964). "Regional Problems and Location of Industry Policy in Britain". Urban Studies, Volume 1, Number 2, 1964.

NEVIN, Edward (1966). "The Case for Regional Policy". The Three Banks Review, Number 72, December 1966.

NORTHERN ECONOMIC PLANNING COUNCIL (1966). "Challenge of the Changing North". London: H.M.S.O., 1966.

NORTHERN PENNINES RURAL DEVELOPMENT BOARD (1970). Annual Report 1970 . London: H.M.S.O., 1970.

NORTH WEST ECONOMIC PLANNING COUNCIL (1968). "Strategy II: The North West of the 1970s". London: H.M.S.O., 1968.

NORTH WEST ECONOMIC PLANNING COUNCIL (1966). "An Economic Planning Strategy for the North West Region". London: H.M.S.O., 1966.

ODBER, Alan J. (1970). "Policy After Hunt". Urban Studies, Volume 7, Number 2, 1970.

ODBER, Alan J. (1965). "Regional Policy in Great Britain". Area Redevelopment Policies in Britain and the Countries of the Common Market. Economic Redevelopment Research, Area Redevelopment Administration, U. S. Department of Commerce. Washington: U.S.G.P.O., 1965.

ODBER, Alan J. (1959). "Local Unemployment and the 1958 Act". Scottish Journal of Political Economy, Volume 6, Number 3, 1959.

ODBER, Alan J., ALLEN, E., and BOWDEN, P. J. (1957). "Development Area Policy in the North East of England". Newcastle-upon-Tyne: North East Industrial and Development Association, 1957.

PETERSON, A. W. (1966). "The Machinery for Economic Planning: Regional Economic Planning Councils and Boards". Reprinted from Public Administration, Spring 1966.

POSNER, Michael (1969). "Regional Economic Policy in the United Kingdom" The New Atlantis, Volume 1, Number 1, 1969.

PULLEN, M. J. (1965). "Regional Development in the United Kingdom". Planning and Growth in Rich and Poor Countries. Birmingham and Ford, Editors. New York: Praeger, 1965.

ROBERTSON, Donald J. (1965). "A Nation of Regions" Urban Studies, Volume 2, Number 2, November 1965.

ROBERTSON, Donald J. and GRIEVE, Sir Robert (1964). The City and the Region. University of Glasgow Social and Economic Studies Occasional Papers Number 2. Edinburgh: Oliver and Boyd, 1964.

RODWIN, Lloyd (1970). Nations and Cities: A Comparison of Strategies for Urban Growth. Boston: Houghton Mifflin Co., 1970.

ROLL, Sir Eric (1966). "The Machinery for Economic Planning: The Department of Economic Affairs". Reprinted from Public Administration, Spring 1966.

ROYAL COMMISSION ON LOCAL GOVERNMENT IN ENGLAND (1969). Report. (The "Redcliffe-Maud Report") Cmnd 4040. London: H.M.S.O., 1969.

ROYAL COMMISSION ON LOCAL GOVERNMENT IN SCOTLAND (1969). Report. (The "Wheatley Report"). Cmnd 4150. Edinburgh: H.M.S.O., 1969.

ROYAL COMMISSION ON THE DISTRIBUTION OF THE INDUSTRIAL POPULATION (1940). Report. (The "Barlow Report") Cmnd 6153. London: H.M.S.O., 1940.

SCOTTISH COUNCIL (DEVELOPMENT AND INDUSTRY) (1961). "Inquiry into the Scottish Economy 1960-61". (The "Toothill Report") Edinburgh: Scottish Council (Development and Industry), 1961.

SCOTTISH DEVELOPMENT DEPARTMENT (1963). "Central Scotland: A Programme for Development and Growth". Cmnd 2188. Edinburgh: H.M.S.O., 1963.

SCOTTISH OFFICE (1935). "Commissioner for the Special Areas in Scotland: Report for the Period 21 December 1934 - 30 June 1935". Cmnd 4958. London: H.M.S.O., 1935.

SCOTTISH OFFICE (1936a). "Commissioner for the Special Areas in Scotland: Report for the Period 1 July 1935 - 31 December 1935". Cmnd 5089. London: H.M.S.O., 1936.

SCOTTISH OFFICE (1936b). "Commissioner for the Special Areas in Scotland: Final Report of Sir Arthur Rose". Cmnd 5245. London: H.M.S.O., 1936.

SCOTTISH OFFICE (1937). "Report of the Commissioner for Special Areas in Scotland". Cmnd 5604. London: H.M.S.O., 1937.

SCOTTISH OFFICE (1938). "Report of the Commissioner for Special Areas in Scotland". Cmnd 5905. London: H.M.S.O., 1938.

SCOTTISH OFFICE (1964). "Development and Growth in Scotland 1963-64" Cmnd 2440. Edinburgh: H.M.S.O., 1964.

SCOTTISH OFFICE (1966). "The Scottish Economy 1965 to 1970: A Plan for Expansion" Cmnd 2864. Edinburgh: H.M.S.O., 1966.

SCOTTISH OFFICE (1971). "Reform of Local Government in Scotland" Cmnd 4583. Edinburgh: H.M.S.O., 1971.

SECRETARY OF STATE FOR LOCAL GOVERNMENT AND REGIONAL PLANNING (1970). "Reform of Local Government in England" Cmnd 4276. London: H.M.S.O., 1970.

SELF, Peter (1967). "Regional Planning in Britain: Analysis and Evaluation" Regional Studies, Volume 1, Number 1, 1967.

SELF, Peter (1964). "Regional Planning in Britain". Urban Studies, Volume 1, Number 1, 1964.

SHONE, Sir Robert (1966). "The Machinery for Economic Planning: The National Economic Development Council". Reprinted from Public Administration, Spring 1966.

SMITH, Peter M. (1966). "What Kind of Regional Planning -- A Review Article" Urban Studies, Volume 3, Number 3, 1966.

SOUTH EAST ECONOMIC PLANNING COUNCIL (1967). "A Strategy for the South East". London: H.M.S.O., 1967.

SOUTH EAST JOINT PLANNING TEAM (1970). "Strategic Plan for the South East: A Framework". London: H.M.S.O., 1970.

SOUTH WEST ECONOMIC PLANNING COUNCIL (1967). "A Regiona with a Future: A Draft Strategy for the South West". London: H.M.S.O., 1967.

STEELE, D. B. (1964). "New Towns for Depressed Areas". The Town Planning Review, Volume XXXIV, 1963-64.

THOMAS, Ray (1969). "Aycliffe to Cumbernauld: A Study of Seven New Towns in their Regions". P.E.P. Broadsheet 516, Volume XXXV, December 1969.

TURNBULL, Phipps (1967). "Regional Economic Councils and Boards". Journal of the Town Planning Institute, Volume 53, Number 2, 1967.

WELSH OFFICE (1971). "The Reform of Local Government in Wales: Consultative Document". Cardiff: H.M.S.O., February 1971.

WEST MIDLANDS ECONOMIC PLANNING COUNCIL (1967). "The West Midlands: Patterns of Growth". London: H.M.S.O., 1967.

WEST MIDLANDS ECONOMIC PLANNING COUNCIL (1965). "The West Midlands — A Regional Study". London: H.M.S.O., 1965.

WILSON, Thomas (1967). "Finance for Regional Industrial Development". Three Banks Review, September 1967.

WILSON, Thomas (1965). "Papers on Regional Development" (Editor). Supplement to the Journal of Industrial Economics. Oxford: Basil Blackwell, 1965.

WILSON, Thomas (1964a). Policies for Regional Development. University of Glasgow Social and Economic Studies Occasional Papers Number 3. Edinburgh: Oliver and Boyd, 1964.

WILSON, Thomas (1964b). Planning and Growth. London: Macmillan, 1964.

YORKSHIRE AND HUMBERSIDE ECONOMIC PLANNING COUNCIL (1966). "A Review of Yorkshire and Humberside". London: H.M.S.O., 1966.

* * * *

June 2, 1971

Notes on a National Urban Development Strategy
for the United States:
Politics and Analytics

Lowdon Wingo*

NOTE: This paper has been prepared for the Resources for the Future-University of Glasgow Conference on Economic Research Relevant to National Urban Development Strategies, Glasgow, Scotland, August 30-September 3, 1971. It is subject to revision and should not be cited or quoted without express permission of the author.

*Director of Urban and Regional Studies
Resources for the Future, Inc.
1755 Massachusetts Avenue, N. W.
Washington, D. C.

Lowdon Wingo is Director of Regional and Urban Studies at Resouces for the Future, where he has been a member of the Research Staff since 1957. He did his graduate work at the University of Chicago and as a Littauer Fellow at the Graduate School of Public Administration at Harvard University. He is editor of the collection Cities and Space, and with Harvey Perloff, of Issues in Urban Economics. Mr. Wingo is author of Transportation and Urban Land and of numerous articles and reports on urban and regional development, metropolitan planning, urban transportation, recreation, urban renewal, and related urban problems. His present research interests include problems of urbanization in developing countries, urban public economics, the economics of national settlement patterns, and the governance of urban systems. He has been a consultant to the Organization of the American States, the United Nations, the World Bank, The Organization for Economic Cooperation and Development and various domestic agencies on matters of urban and regional development.

INTRODUCTION

The purpose of this paper is to explore the content and implications of a set of proposals advanced by public figures and private parties the United States under the rubric of "national urban growth policies." Unfortunately, national urban growth policy is not a technical term whose content is precisely defined; instead, it embraces a number approaches or strategies which have been put forward in recent months to cope with the general problems of an urban society. While there is reason to believe that no political consensus is to be found which would promise early adoption of such policies by the federal government, it is likely that the growing urgency of the problems of the major metropolitan areas in the United States and increasing awareness of the distress of nonmetropolitan America will energize a continuing search for a general national strategy which can make coherent the impacts of the vast array of measures which a society solarge and so productive adopts to maintain its viability. It is this prospect which justifies a continuing examination of the diagnoses and prescriptions which will be offered to set things right in the nation. It is to the advancement of this examination that this paper is addressed.

The task of policy formation in a representative system is to discover a prescription which is both consistent with the etiology of the social disorder and sufficiently palatable to the patient to assure his continuing acceptance. Analytical realism and political feasibility are the necessary attributes of any public policy and certainly no less so for urban growth policies. The scholar cannot always tell the successful politician much about political feasibility, but he can -- and must -- constantly appraise the extent to which policy proposals are realistic in terms of the analytical characteristics of the problem. He does this by asking two kinds of questions: does the positive model underlying the policy correspond to the real world? If so, are the means chosen well calculated to achieve the ends? This paper proposes to explore some aspects of the positive and normative models implicit in the current urban growth policy discussion.

It will seek to lend content to this concept by first examining what the advocates -- politicians, bureaucrats, and movers on the urban scene -- consider to be the meaning of urban growth policy, and then by examining some of the essential analytical questions which seem to be implied by the concept. At the end, it will speculate briefly on how to relate the policy and the analytical images of national urban development strategy.

American Urban Development Policy: A Very Modern History

Public Law 91-609 became law of the land on the last day of 1971 at the very end of the 2nd Session of the 91st Congress. Its preamble reads, "An Act to provide for the establishment of a national urban growth policy, to encourage and support the proper growth and development of our States, metropolitan areas, cities, counties, and towns with emphasis upon new community and inner city development, to extend and amend the laws relating to housing and urban development and for other purposes," it is otherwise known as the Housing and Urban Development Act of 1970.[1]

However general the Preamble, signature of this act by the President put the federal government officially into the business of defining a national interest in the evolving characteristics of the settlement pattern of the United States. Such assertion, however, conceals more than it reveals: The passage of the act was a nominal step at best toward a more comprehensive and articulate role of the national government in subnational development in this country. It is no secret that the Nixon Administration vigorously opposed the inclusion of Title VII, the so-called Urban Growth and New Community Development Act of 1970, in this legislation as being out of joint with its own legislative program for the cities, but in the end skillful maneuvering by liberal forces in a Democratic party firmly in control of both houses of a Congress impatient with the Executive branch allowed the Congress to seize the policy initiative from the President and force its will upon him. From a substantive point of view the victory is hollow: it imposes on the President a single responsibility, "to transmit to the Congress, during the month of February in every even numbered year beginning with 1972, a Report on Urban Growth for the preceding two calendar years, which shall include [in addition to information about the state of urban development] recommendations for programs and policies to carry out such a policy . . ."[2] Indeed, the impact of Congress' concern about the character of U. S. urban development on the President's point of view is reflected in a single statement in the President's major policy pronouncement on federal policy for the cities:

> "The concentration of population growth in already crowded areas is not a trend we wish to perpetuate. This administration would prefer a more balanced growth pattern -- and we are taking a number of steps to encourage more development and settlement in the less densely populated areas of of our country."[3]

The only "step" apparent in this statement is scarcely dramatic. After reserving 80% of the proposed $2 billion funding for the program for distribution by formula to SMSA's via the states, and after some proportion of the remaining 20% has been distributed to the same SMSA's to protect them from loss resulting from the changeover of policies, the remainder would be available to the Secretary of Housing and Urban Development to distribute "to encourage state involvement in urban community development, to perform research, to demonstrate new techniques and to aid localities with special needs and with special opportunities to implement national growth policy."[4] In short, under the best of circumstances as interpreted by the proponents of national urban growth policy, for every dollar that the administration would spend to reorder the settlement pattern of the country it would spend roughly ten to support and improve the existing settlement pattern and population distribution under the Special Revenue Sharing Program for Urban Community Development and Planning.

Thus, today two views of national urban policy have emerged one from a conservative Republic commitment to the Federal system and now dramatized by a Republican president's "New Federalism," the other from the rapidly increasing involvement of the federal government in urban affairs under the Kennedy and Johnson administrations, a role for which a hard core of congressional liberal Democrats has become steward.

The evolution of this debate on the content of national urban policy has not taken place in a vacuum, although a description of the historical roots of this debate is not only beyond the purview of this paper but has already been well discussed and documented by Lloyd Rodwin in his study of national urban policies around the world.[5] The United States has in the postwar period evolved a powerfully effective, implicit national urban policy whose outcome is a compound of policies, programs and plans addressed to other, quite explicit public objectives at every level of the federal system. A key element in this implicit policy is thirty years of a national agricultural development policy which gave the U. S. the most highly capitalized and most productive agricultural establishment in the world while transferring roughly fifty million people, one fourth of the current national population, from the nation's rural farm areas to its cities and suburbs.[6]

Hardly less significant if more recent have been the civil rights policies and the war on poverty, which made the nation's great urban centers the foci of these political struggles and the economic meccas of the deprived and disadvantaged groups in American society.

Add to these policy developments a National Defense Highway program, which opened up enormous land development reserves in the hinterlands of every major city, and a national housing policy which made an American ideal of the free-standing, single-family dwelling on its own parcel of ground, and the stage is set for the low-density suburban explosion, which has been a major outcome of American policy. The other side of the coin is, of course, the residual community which the suburbanizing middle class left behind, the excluded minorities, the poor, the old. They were left behind because it cost money to join the suburban "club" -- to move, to pay rents for new housing produced by an industry which never learned how to produce low cost housing, to pay taxes, to pay transportation costs to jobs and public services. Thus, the built-in exclusionary principle imbedded in these policies acted like a filter sorting out the affluent from the poor, and where these implicit processes were not sufficiently fine-grained, the federal system made possible finer screens: "home-rule" communities surrounding the central city had only to enact ordinances whose effect if not explicit purpose, was to exclude. Aided by the civil strife of the late sixties, our implicit national urban policies were producing a newly segregated society in which the battle line of an increasingly acrimonious class confrontation are the boundaries of the central cities of our metropolitan regions, a condition described by the report of the Kerner Commission in 1968.[7]

Both Kennedy and Johnson Administrations attempted to ameliorate these developments not by reversing the policies but through enactment of a multitude of programs addressed to specific problems thrown up by these developments. The key element of their strategy was the federal grant-in-aid to qualified local communities for specific projects and programs. "Qualified communities" were those who could give evidence of their own participation in the solution to these problems in the form of "workable programs" and "overall economic development plans." In the eight years of these administrations a large number of such categorical grant programs were enacted, the incoming Nixon administration charging that the condition of cities was worse at the end of that period than at the beginning. And thus the New Federalism!

The issue itself is so recent in the American scene that neither party is certain that it makes good politics. Indeed, one informed observer has complained publicly that in the months following the President's State of the Union Address in January of 1970 the administration's interest in national urban policy so flagged that it had become lost in the White House staff.[8] As the Johnson Administration turned its affairs over to the Republicans in January 1969 two major commission reports and major private policy study were circulating among government agencies, none of which gave any recognition to the need for national intervention in the development of the national system of cities. The Brookings Institution's thoughtful Agenda for the Nation dealt with poverty, racism, education, law enforcement, and housing, but nowhere suggested that these and other problems identified with cities could be ameliorated by some general national policy specifically addressed to the cities.[9] The National Commission on Urban Problems (Douglas Commission) in its Report Building the American City recommended a number of ways in which federal policies could expedite the development of urban areas across the nation, but nowhere suggested any overarching federal policy about urban growth.[10] The narrower mission of the President's Committee on Urban Housing (Kaiser Committee) is reflected in its report A Decent Home, which treats housing as a national problem without suggesting that its solution can be influenced a lot by changes in population distribution or by managing the location of housing and new job-creating capital over the national landscape.[11]

Indeed, the recent roots of this policy concern are to be found, mirabile dictu, in the office of Johnson's Secretary of Agriculture Oville L. Freeman. In an attractive if polemical booklet Communities of Tomorrow the Department of Agriculture was to be found busily staking its claim to problem piece of urban America, the nonmetropolitan areas and their urbanization trends and problems:

> " . . . our metropolitan areas have more people and problems than they can cope with. All around us they are exploding with violence. At the same time, many villages, small towns, and their surrounding countryside are being drained of people and economic vigor . . . We intend to help people build [Communities of Tomorrow]. Our objective is to remove the scars of collision between man and his environment and to avert further collisions . . ."[12]

A month later Freeman flanked by five other Cabinet members opened a national symposium on "Communities of Tomorrow -- National Growth and its Distribution" in Washington, saying

> "This symposium was born at a luncheon held by its Cabinet sponsors a few months ago. All of us expressed concern at the continuing exodus from countryside to big city and the problems that it helped to create and that it is helping to perpetuate. We asked ourselves if it made sense to compress more and more people into less and less space, and, in pondering the answer, we had to agree that the country had never really addressed itself to the question."[13]

Secretary Weaver of HUD underneath the proper courtesy and cordiality attending such things treated Agriculture's initiative somewhat as claim-jumping in his cautionary remarks:

> " . . . the increasing concentration of population in the great metropolitan centers is a phenomenon being experienced in all of the nations of the world. Whatever our feelings may be about this trend, there is no evidence available that it is reversible . . ."
>
> "When we speak of fostering balance within and between regions, I do not believe that we should expect to reverse the trends of industry location within metropolitan areas. The markets, the business services, and the institutions that industry requires all exist in these areas."
>
> "Finally, I believe that as we look ahead for several decades we cannot expect an indefinite continuation of a high level of rural to urban migration."
>
> "For we must recognize that regardless of what happens to patterns or volumes of migration, the great increase in jobs will be in non-agricultural industries and the great increase in population will be in urban America."[14]

It is clear that the idea of a national urban growth policy generated greater enthusiasm outside of the key agencies in the executive branch concerned with urbanization than within, a situation not greatly different today.

Perhaps the most important single document in the evolution of policy interest in national urban growth policy was issued by the Advisory Commission on Intergovernmental Relations in April of 1968. Urban and Rural America: Policies for Future Growth took the entire issue out of the political forum and gave it intellectual as well as rhetorical credibility.[15] Analysing recent history of urbanization and subnational economic growth in the U. S. and weighing the pros and cons of a larger federal role in urban development, the Commission concluded:

> " . . . taking particular account of its findings as to diseconomies of urban congestion, the locational mismatch of jobs and people, the linkage of urban and rural problems, and urban sprawl -- that there is a specific need for immediate establishment of a national policy for guiding the location and character of future urbanization, involving federal, state, and local governments in collaboration with the private sector of the national economy. The commission's findings further suggest that such a policy would call for influencing the movement of population and economic growth among different types of communities in various ways so as to achieve generally a greater degree of population decentralization throughout the country and a greater degree of population dispersion within metropolitan areas."

Not only did the ACIR Report identify the substance of a national urban growth policy, it went on to describe the elements of such a policy calculated to deal with industrial location, population mobility, and new community development, and to discriminate the roles of federal, state, and interstate planning agencies. It advanced the debate from rhetorical generalities to policy particulars and armed its proponents with an impressive array of facts, figures, and analysis.

In May of 1968 the theme was picked up in the forum of the national Congress. Senator Fred Harris of the Government Operation Committee of the Senate presided over The National Manpower Conference in Stillwater, Oklahoma, addressed to "The Rural to Urban Population Shift: A National Problem" saying:

"Despite the increased effort and investment in programs to assist cities, the problems each year become greater; more and more people continue to pour into the cities from the countryside, induced in part, by the very programs designed to solve the problems of the cities. . . . In order to stabilize our rural areas and small town population, we must foster if we can, a rethinking of national policy . . . efforts to solve the problems of the cities must be accompanied by innovative and imaginative efforts to provide greater opportunity and viability in rural areas, smaller towns, and smaller cities -- and new cities.[17]

Thus, in the last year and a half of the Johnson Administration a theme that the President broached at Dallastown, Pa., in September of 1966 -- "The cities will never solve their problems unless we solve the problems of towns and smaller areas. So consider the problem of urban growth" -- was finding acceptance in key Washington political institutions.[18] The Nixon administration, entering office in January 1969 fell heir to the issue without any specific commitments concerning it, except those implicit in its Democratic party origins.

In May of 1969 conflicting concepts of national urban policy appeared in two public statements by key members of the President's Urban Council. Vice President Agnew wrote the introduction to the report of the ad hoc National Committee on Urban Growth Policy, which took a quick look at future demographic dimensions of America's future and recommended a massive new federal program aimed at the construction of 100 new towns for populations of 100,000 each and ten new cities of a million each.[19] The Committee report was notable in that it was set up under the sponsorship of major urban political interest groups-- the National Association of Counties, the National League of Cities, the United States Conference of Mayors, and Urban America, Inc. -- who were apparently willing to accept the beguiling simplicity of this report and its recommendation to be the basic element of national urban policy.[20]

Several days earlier, however, White House urban adviser Daniel P. Moynihan speaking at Syracuse University laid out his now notable ten points for an urban policy, but the language and concepts were very different from those employed either by the Democrats before him or by the Vice President and members of the National Committee on Urban Growth Policy. Pointing out that the problems of cities had continued to grow in spite of the fact that federal domestic programs had grown from 45 to over 400 during the preceding two administrations, he concluded that the instrument of categorical grants to carry out a multiplicity of government programs had failed to create the powerful incentive for governments and private bodies to behave in a manner consistent with the general interest, and that so many programs addressed to so many problems ignored the basic and complex interdependencies among urban problems and processes. What is needed, he argues, is not so much 400 programs as a national urban policy which recognizes the interrelatedness of the urban system and maximizes the capacity of the conventional units of general government to deal with the problems as they are seen not by Washington bureaucrats but by the constituents of the local governments affected.[21]

Behind Moynihan's ten points is to be found two major themes which define the federal interest in the cities. The first is the need for a transformation of a dependent poverty class inherited from preceding administrations into a stable, independent community enjoying equal access to economic opportunity and public goods with the other classes of the society.

Housing, education, welfare, employment opportunities must be accessible everywhere in the metropolis, and especially to the citizen who is willing to "vote with his feet in choosing the best constellation of opportunities for himself and his family." The second major theme is the reform of those parts of the federal system which obstruct the ability of states and local areas to deal with their problems. This involves the strengthening of local general governments to cope with their purely internal problems, the reduction of "paragovernments" addressed to special problems but having no political responsibility to a well defined, affected constituency. It involves a new relationship with the federal government as the setter of broad national guidelines and as a source of funds both to amplify the effects of local policies and to reduce the disparities in those social services having the most to do with economic opportunity.

In addition, Moynihan concerned himself with the development and protection of urbanity, of the special qualities of life promised by the potential of the great cities. The environment and other natural resources, aesthetic qualities, cultural vigor, and social amenities are all appropriate targets for urban policies, but on the whole, in Moynihan's estimation, they add up to something much less critical than the need to provide full citizenship in the economy to the poor and the insistent demand of the times to reform and energize the federal system anew.

Moynihan's urban policy, hence, is not really concerned with the national settlement pattern nor with population distribution, nor with interregional social welfare disparities. It is concerned with structure and processes, the automatisms of society which can and should provide the compensatory mechanisms for the adjustment of "imbalances" in the urban system. Implicitly, as he distrusts the capacity of 400 or 800 federal programs to resolve the national urban problem, he finds suspect the concept of a settlement pattern policy:

> " . . . in the next 30 years we shall add one hundred million persons to our population. Knowing that, it is impossible to have no policy with respect to where they will be located. <u>For to let nature take its course is a policy</u>. To consider what might be best for all and to seek to provide it is surely a more acceptible goal."[22]

While Moynihan spoke in Syracuse as a private person, the authority of his statement was underlined in February of the following year when the President used virtually identical language in his address to the Annual meeting of the United States Conference of Mayors meeting in Indianapolis.[23] The Moynihan diagnosis and prescription to the problem of the cities thus prevailed over those of the Vice President, and the idea of a national settlement pattern policy under this administration appeared to have disappeared into that repository of beguiling ideas weighed on the complex scales of politics and found wanting.

The most recent chapter in the political-intellectual history of the concept of a nationa urbanization policy is to be found in the President's special message to Congress on Special Revenue Sharing for Community Development and Planning in which the failure of the federal system to deal with the urban problems is diagnosed as being due to fragmentation and federal control:[24]

> "[Given the multitude of federal programs] it is extremely difficult for any community to create an overall strategy for development when each element in that strategy must be negotiated separately by officials who cannot be surer about the outcome of all the other negotiations."

> Morever, "Decisions about the development of a local community should reflect local preferences and meet local needs. No group of remote federal officials -- however talented and sincere -- can effectively tailor each local program to the wide variety of local conditions which exist in this highly diversified land. The only way that can be done is by bringing more tailors into the act, tailors who are elected to make sure that the suit fits the customer."

Finally, the general prescription:

> "The genius of the federal system is that it offers a way of combining local energy and local adaptibility with national resources and national goals. We should take full advantage of that capacity as we address the urban challenge."

The Nixon proposal would consolidate four major urban categorical grant programs into a single fund of roughly $2 billion the first year to be joined in the second year by appropriations to the Community Action Program of the Office of Economic Opportunity. Eighty per cent of these funds would be allocated by formula weighted by need factors to the 247 SMSA's and their component jurisdictions to be used for any purpose reasonably related to community development. The remaining twenty percent would be allocated at the discretion of the Secretary of HUD for a number of purposes enumerated specifically. In a later message, the president proposed to reorganize the domestic departments of the Executive branch to align them with his six special revenue sharing programs the more easily to relate "national resources and national goals" to local energies and adaptability."[25]

Nevertheless, settlement policy is not entirely dead in the policy councils of this administration; it has reappeared under another name -- population distribution policy. The report of the President's National Goals Research Staff published in July of 1970 addressed among other things federal policy with respect to population growth and its distribution.[26] The particular style adopted by the report was to amplify the "debates" between policy alternative in critical areas for the deliberation of the President without taking a stand. However, the text rarely leave the reader in doubt about the preferences of the Staff, and in the matter of population distribution the staff seems to line up squarely with the desirability of explicit national policy on population distribution.

Indeed, it virtually adopts by reference the ACIR Report, Rural and Urban America and so allies itself strongly with the substance of the concept of urban growth policy which had evolved under the Johnson Administration:

> "[Recent] research . . . suggests that trends towards megalopolis in some areas and under-population in others are reversible. It also suggests there is an opportunity for a different and more rewarding future for the Nation as a whole, than the discouraging vision of gargantuan megalopolis and rural desolation. But realization of a better future will probably require a coordinated national strategy for balanced population distribution. The federal government can provide leadership in developing such a strategy, but public and private institutions across the country will need to participate in both planning and implementation."

> "[In contrast to the choice of "no change in public policy"] there is the choice of a decisive public policy and action to achieve a more promising future for the country as a whole. The objective of this choice might be to promote more balanced demographic growth in order to affect positively the quality of life in both urban and rural America.[28]

Proceeding to examine three basic strategies suggested by experts--population spread by generating growth in sparsely populated rural areas, concentrating growth in alternative growth centers in nonmetropolitan areas, and construction of new cities outside the large metropolitan regions -- the Goals Staff concluded:

> " . . . we need to decide on whether or not we will adopt a deliberate strategy to encourage internal migration to negate the forecasts of ever-growing urban congestion in a few megalopoli (sic). A viable option for such an alternative strategy is a policy of encouraging growth in alternate growth centers away from the large urban masses, coupled with a complementary effort for the use of new towns.[29]

These speculations on population policies by the National Goals Staff anticipated the investigations of the Commission on Population Growth and the American Future, which had been established by Congress by statute and signed into law by the President on March 16, 1970. It is the Commission's responsibility to study in depth among others the issues of population distribution on which the Goals Staff made its poorly concealed recommendations, and in March of 1971, as it approached the midpoint of its statutory life, the Commission issued an Interim Report.[30] Pointing out the 70% of the 1960-1970 population growth in metropolitan areas occurred as a result of natural increase and the migration contributed less than a third, the Commission commented that

> ". . . simply redistributing population might not solve many of the population-related problems we face.
>
> Moreover, a large population such as ours might not be able to live at its present standard of consumption without high concentrations of people and economic activity . . . Even so, it may be desirable to slow or stop the growth of very large metropolitan areas . . . it would be very difficult to do this without slowing the growth of total population."

The cautious tone of this pronouncement is in keeping with the state of completion of the Commission's task and also with the complexities, political, economic and social, which in fact surround implementation of policies to alter substantially the American pattern of settlement. How the Commission will finally conclude on this topic will not be known for another year, but in this its first public statement it has established a properly circumspect intellectual and political style.

The Nixon administration has spelled out in large degree its own urban policy for the nation centered around the reform of the structure and process of the American system. Until Congress accepts the proposed policies, putative national urban policy is simply the working out of laws, programs, administrative decisions and regulations, and judicial rulings which have accumulated over the past two generations. The pressure for the administration to take a more concrete position on population distribution and settlement pattern policies is coming from a number of directions -- from a Democratically controlled Congress, from major urban pressure groups, from independent commissions, from within the executive family. For the President to win this contest, the New Federalism will have to have a chance to pay off. Such an outcome will be hard to come by. The next election is only months away and Congress has not shown itself sufficiently interested in the proposals to push them forward rapidly. Hence, while the prospects for the adoption of an explicit national urban growth policy of any persuasion in the near future do not appear to be very good, the discussion will obviously continue and elements of a more wholistic national policy for urban growth will be tested for analytical realism and political appeal, which justified a continuing appraisal of policy instruments and the models which relate them to the desired ends and goals of national policies.

Nonmetropolitan America in a National Urban Growth Policy

The development pattern of nonmetropolitan America is more strongly characterized by poverty and economic distress than that of the metropolitan areas. Taking the percent of the population who fall below the poverty line as a negative welfare measure, the incidence of poverty was two and a half times greater in nonmetropolitan than in metropolitan areas in 1965, a disparity even greater than that between rural and urban areas (one and two thirds). At the extremes, the incidence of poverty in rural farm areas was almost five times that in the suburban rings of the metropolitan areas:

	Percent poor	Index of Poverty[a] (U.S. = 100)
Metropolitan Areas,	12.6	67
(Suburbs)	(6.7)	(33)
Nonmetropolitan Areas[b]	32.0	150
(Rural farm)	(29.7)	(193)
Urban Areas	14.8	81
Rural Areas	29.7	155

Notes: a. $\frac{\%\text{ poor in sector}}{\%\text{ nonpoor in sector}} \div \frac{\%\text{poor in nation}}{\%\text{nonpoor in nation}}$

b. Nonmetropolitan areas include rural areas plus urban areas outside of SMSAs.

SOURCE: The President's National Advisory Commission on Rural Poverty, The People Left Behind, A Report, (Washington, USGPO, 1967)

If nonmetropolitan poverty were generally diffuse, the settlement pattern issues would hardly arise, but the data above suggest some substantial disparities among the rural-farm, rural-nonfarm, and urban areas. These issues are highlighted by recent tabulations by Clawson showing that forty-five percent of U.S. counties lost population in the 1960-1970 period and two-thirds of those had lost population in one of the two preceding census periods. Only one third of the 3,072 American counties tabulated showed consistent gains in population over the three decades. Add to these observations the fact that 57% of SMSA counties have been steady gainers throughout the twentieth century as against 9% of the nonmetropolitan counties, and the general dimensions of the growth problems of nonmetropolitan American become somewhat clearer.

Legislative policy in the U.S. toward the problems of rural poverty has been a combination of two not completely compatible approaches. The area development strategy has been addressed to counties exhibiting substantial disparities from national averages in income or unemployment, mainly by making available loans on favorable terms to new private enterprise which promises to reduce unemployment and by issuing loans or grants to local government units to improve their basic infrastructure. The Economic Development Administration is currently responsible for this program. The Appalachian Regional Commission pioneered the approach to nonmetropolitan development through the coordination of state and federal programs in large multistate regions. Using a growth center strategy, the Appalachian Regional Commission attempted to mobilize development infrastructure, concentrate human resources development programs, and upgrade and improve access to improved public services. At the end of its first five years it was able to report some overall improvement in its region measured by some closure of the income and employment gaps between the region and the rest of the country, although outmigration continued at a high rate.[31] The Appalachian model has been replicated in the Title V regional commissions set up under EDA for five other such statewide or multistate regions, but their performance has been unspectacular to date.

Nevertheless the poor performance of the nonmetropolitan economy continues to trouble the nation and has resulted in a number of policy recommendation in recent years. The report of the President's National Advisory Commission on Rural Poverty in 1967 took an across-the-board approach focusing on human resource development to complement broad national economic policies, but disregarding the issues assocated with the inadequacies of the nonmetropolitan settlement pattern.[32] The Republican Administration appointed its own task force, which brought forward its report in 1970, in which the emphasis shifted toward (1) reorganization of government programs along the general lines of Nixon's New Federalism and (2) some of the settlement pattern issues by calling for "a national policy on the geographic distribution of population and economic growth" and a development strategy based on growth centers.[33] Much of the content of both reports was passed over, however, in the policy subsequently adopted by the administration and expressed in the President's Message to Congress on Special Revenue Sharing for Rural Development, the main thrust of which w to take the Federal government out of the distressed area picture in every respect except for augmenting state and local budgets from the federal tax take.[34] The regional commissions would have been dismantled as well as the national area development agency, and their coordinative role would have been replaced by state development plans to be filed in Washington to qualify for the federal funds. Congressional displeasure was demonstrated when, within a fortnight, the Senate approved extension of the Appalachian Regional Development Act for four years by a vote of 77-3! Thus, broad rural development policy in the U. S. is very much up in the air.

The development of rural and nonmetropolitan America raises two fundamental questions for national urban growth policy. First, have existing policies and market imperfections distorted the allocation of resources in the national economy in favor of metropolitan America and against the welfare of the citizens of nonmetropolitan America? In the mind of former Secretary of Agriculture, Orville Freeman, the popularizer of the concept of "rural-urban balance," there was no doubt about this. Second, <u>within</u> nonmetropolitan America, are there alternative distributions of population, infrastructure, and economic activity that would improve the welfare of that sector of the national economy. Clearly these two questions are not completely severable, but the policy issues raised take us different directions.

The nonmetropolitan sector has traditionally carried out the function of servicing the primary sectors of the national economy and hence has always been sensitive to secular developments in agriculture and mining. The rapid technological change experienced by both these sectors in the last generation has supplanted millions of farmers and miners and casual workers with new techniques, machines, and processes, as well as with more highly capitalized labor. Although the speed of these changes has exaggerated the disequilibria of adjustment, the changes themselves are clearly consequences of the economic developed processes with deep roots in American economic history. Nevertheless, the rapid contraction of jobs in farming and mining and the declining share of labor in the value added in those two sectors has severely eroded the economic base of the network of villages, towns, and small cities which had grown up in the preceding century to service the primary sectors of the economy. The resulting disinvestment and the shrinkage of secondary employment in these communities has been accompanied by a complementary growth of net outmigration toward the metropolitan areas.

These shifts of the nonmetropolitan labor force toward metropolitan employment exhibit high degree of selectivity, which tends to be exaggerated in times of rapid change. Some are left behind, but the working age population which migrates from a community tends to be younger, better educated, more adaptable, probably more ambitious, and has less sunk investment to protect than those who stay behind. The latter consequently tend to be less productive, less mobile, more attached to their surroundings, and more dependent, on the whole. Hence, nonmetropolitan American presents a contrast of prosperity and growth in the reorganized, technologically sophisticated primary sectors themselves with the traditional enterprises and the supporting network of settlements. It is this rural poverty and the distress of the small town in nonmetropolitan America which is frequently cited as evidence of rural-urban imbalance, that is, of an allocation of resources unfavorable to nonmetropolitan America.

The hypothesis implies that mining and agriculture are treated unfavorably in national factor markets. Yet price supports, agricultural surpluses and depletion allowances suggest that the primary sectors consume more than what would be their share of resources under more perfect market conditions. However, the sizes of these sectors have long been determined by national policy rather than by the hidden hand of the market. The continuing outflow of labor and capital from nonprimary sectors in nonmetropolitan America toward more profitable metropolitan employment suggests that the lag in response to their diminished economic base results in their being too large, also. Hence, in the matter of metropolitan-non-metropolitan balance the evidence does not clearly point to the conclusion that the national welfare would be improved by arresting these factor flows.

The effectiveness of the distribution of population and economic activity within non-metropolitan America is another question. The rapid changes in agriculture and mining in the past generation have left the nonmentropolitan settlement pattern in a state of disequilibrium, accentuated by the growth of the national highway system and the widespread use of the automobile which have given many rural people a new mobility as consumers that is not reflected in the settlement pattern or even in the distribution of public services. Where local farms used to be served by village stores and services a few miles away, now these are being superceded by larger centers serving populations within radii of fifty miles or more with a richer array of goods and services than was possible a generation ago. It is not surprising that the small towns and villages in nonmetropolitan American were the most dramatic losers in the 1970 Census, while the small city tended to be the gainer as it took over the role previously played by more widely scattered small settlements.

This evolution is still incomplete, however. As a result, substantial disparities in economic welfare are to be found among the cities and towns outside the metropolitan areas. The slowest element to change has been the public service sector, dominated, as it is by the inertia of the structures of state and local governments.[35] A large fraction of the rural population enjoys few public services, and those generally of low quality, in part because public institutions in rural areas are so structured as to make effective local response to the needs and demand for public services impossible. Hence, public service functions appear to be inadequately performed everywhere in nonmetropolitan America but in the cities.

There is, hence, reason to believe that the populations of the rural areas and small communities which grew up servicing them would find their welfare considerably enhanced by a reorganization of the nonmetropolitan settlement pattern which would seek to consolidate economic activities and local consumer and producer services in a vastly reduced number of centers, which could also provide opportunities for delivering public services which could enjoy some greater economies of scale. A carefully restrained number of such centers might permit some to become important employment foci for rural outmigrants and act as staging centers for further migration into the major metropolitan areas.[36]

X X X

The diagnosis that emerges from the recent literature advocating national urban growth policy for the U.S. addresses the entire settlement pattern, urban and rural, and can be summarized in the following points:

(1) The subsidies implicit in the production externalities in the very large metropolitan regions has resulted in an excessive concentration of private investment and public capital in such areas;

(2) this concentration of capital has induced excessive migration not only of prime segments of the labor force but also of others drawn by better public services and superior welfare support to such centers resulting in the sprawl of development over their hinterland and spreading slums and blight in their inner cities;

(3) this flow of migrants toward the large cities has left behind in non-metropolitan America small towns and rural areas deprived of effective labor forces needed to maintain their economic viability, which tends to accentuate the national concentration of capital;

(4) meanwhile, the white middle class and new jobs follow each other to the suburbs, where living costs and exclusionary policies contain the poor and the black in the central city isolated from the new jobs and hence increasingly dependent on public support;

(5) this movement of jobs and whites to the suburbs also represent a movement of fiscal resources away from the central city, so that the growth of inner city dependency is accompanied by shrinkage of the fiscal capacity of the central city precipitating a general financial crisis, which can only be resolved by the intervention of state and federal government;

(6) simultaneously, the increasing concentration of population and economic activity in the metropolitan region generates a growing volume of negative externalities which tax the whole region but no activity to the point where its marginal social costs of an activity are equated to its returns and the result is that all such concentrated activities are either subsidizers or subsidized;

(7) the most menacing of these externalities is the degradation of the physical environment of the large city as the volume of solid, liquid, and gaseous wastes presses upon the absorptive capacity of the environment.

Imperfections in market processes augmented by incoherent public policies have resulted in a disposition of population and economic activities over the national landscape which is the source of many of our rural and urban development problems, so goes the diagnosis. The proponents of national policies to correct their state of affairs fall into two classes, those who would improve the performance of market and political processes, and those who would more directly rearrange the settlement pattern itself through such measures as the creation of the New Cities, restricting the growth of major metropolitan areas, managing the volume and direction of internal migration, and more stringently regulating industrial location. Indeed, much of the difference among points of view held by growth policy advocates reflects the differing orientation planners interested in an idealized end state and social scientists interested in capacity of social processes to achieve socially desirable outcomes.

While this general diagnosis would not be disowned by most advocates, it takes little reading of the literature to emphasize the differences in objectives which have been brought forward. Some free-wheeling content analysis of a sample of seven statements concerned with national urban growth policy reveals some sixteen goal statements which are reducible to a few major themes. A major emphasis is found in the race-internal migration theme, which ranges from a deep concern with racial tensions and violence in the cities to a somewhat more philanthropic interest in the conditions of life of urban minority groups, but exhibiting a general agreement that something should be done to abate the migratory flows of the dependant poor to the major metropolitan centers. The second theme strongly emphasized is the quality of life and ranges over environmental deterioration in large cities, the conservation of metropolitan land and water resources, and the esthetics of the city.

The third theme concerns the welfare effects of the political fragmentation that characterizes American metropolitan areas. Intrametropolitan disparities in fiscal resources among the suburbs and between the suburbs and the central city have become a key target for the financial reform of urban government, especially since there seems to be a built in tendency for these disparities to widen in the absence of any redistributive mechanisms in the public fiscal system.

Finally, urban scale and population concentration come in for some attention. The pervasive, if not coldly analytical, feeling is that the big city is the major culprit in many of our internal development problems and the diminution of its relative importance in the national economy could result in little but good for the country as a whole. Scattered interest in improving the performance of state government in augmenting the viability of small urban centers, and in providing more effective structures for decision making about urban development strategies can be found. Thus it is the urbanization of the poor and minorities, the deterioration of the quality of life, the distributive consequences of urban political organization, and the issues of what George Tolley has labeled "city bigness" that emerge as the sensed problems for national urban growth policy, at least as seen by its advocates.

Public policy has exercised a strong if inadvertent impact on the American settlement pattern within the last generation. It is not beyond belief that equally powerful policies could influence substantially the subnational development of the country during the next generation. Such policy places a high premium on being able to anticipate the impacts of public policy on internal development in the United States, and it is not obvious that the current state of knowledge places us in a strong position to do that. A look at some of the issues in the analysis of urban systems bear this out.

Some Extended Hypotheses on the Organization and Welfare Aspects of the Urban System

In the appraisal of urban growth policies there are a number of essential relationships between characteristics of the urban system and both behavioral variables and normative conditions: we would like to know something about the stability characteristics of the system and changes in welfare attendant on moving to other stable states. Certainly if the system is so stable that no available social interventions can substantially alter it, urban growth policy is academic. If changes in welfare cannot be ascribed to changes in the state of the system, the problem is uninteresting as a policy exercise. We need to know enough, then to build reasonable positive models that describe changes in key characteristics of the urban system as functions of the behavioral propensities of individuals and organizations in a reasonably closed society. And we have to have ways of identifying the relative goodness or badness of alternative outcomes. To meet these conditions with respect to national urban growth policies as the issues have been formulated, it would be useful to know more about the following conceptual issues.

(1) the general analytics of optimum city size

(2) private costs and benefits as a function of city size and economic composition;

(3) negative and positive externalities as a function of city size and economic composition (and vice versa)

(4) the distribution of city sizes in a national system as a function of the composition of the national economy, technology, incomes, consumer preferences and transport costs.

(5) the current and capital costs of producing and distributing overhead services as a function of city size;

(6) the welfare characteristics of internal migration.

This set is not exhaustive, but intuitively I give these questions high priority in modelling the urban system.

(1) Policy relevance of the concept of optimum city size. The classical form of the case for regulating the size of cities has been that beyond some point in the size of urban settlement cities are too large and below some point cities are too small, which suggests that there must be a "best size" somewhere within the range identified by those two points to which all cities should aspire. Indeed, the best of all possible worlds would find urban population divided among cities of equal size, that size being optimal. A great deal of speculation and some desultory investigation have gone into identifying either that optimal size, or some range of city sizes which would include the optimal and whose extremes represent only nominal departures therefrom. Convergence on this size, or this range of sizes, would then be the objective of public policy for the national settlement pattern. Most of the speculation has come to rest on some number or set of numbers in the 100,000 to 1,000,000 range with the lower end of the scale preferred, and 250,000 seeming to be the most popular number.[37] This literature does not always discuss how cities got to be so nonoptimal, but the implication is that the pursuit of profit and personal interest rarely if every leads to outcomes consistent with social interests, that is to say, that the private costs and benefits of locational divisions by firms are not congruent with public costs and benefits.

Alonso's suggestion that optimal size of city may be a function of its location vis-a-vis other large cities implies that this kind of a numbers game is not very fruitful, however appealing the idea of some simple numerical goal may be;[38] and by inference forces us to ask what it is that is being optimized. Indeed, much of the literature is addressed to the cost side of the ledger arguing that the costs of cities to their residents increase as we move either way on the size scale from the optimum-inclusive range. In this fashion the problem is reduced to a cost minimization problem which carries with it the assumption that gross benefits to society are invariant with respect to size of city or the distribution of city sizes. Moreover, most of the literature is not concerned with social costs, but with public costs of providing public services, a still narrower formulation of the problem popular because of the relative ease of measurement but hardly the more valid for that.

Two other characteristics of this approach are worth bringing front and center. In the first place, the hypothesis is single-valued: the net benefits, which are cost savings, will not vary with any other characteristics of cities subject to change over time: one set of numbers will suffice for all cases. Furthermore, the static formulation of the hypothesis makes it difficult to apply to a real world in which there are present clearly described processes of urban growth and change; hence it is curiously inappropriate as a public policy concept. The partial, static, single-valued definition of optimum city size simply proposes that a specific city at a particular point in time might improve the welfare of its residents by growing either larger or smaller. There is nothing in the concept which necessarily requires that the same proposition holds for the city at other points in time or for other cities of similar size at the same point in time. The optimum size of city concept, hence, provides us with no capacity to deal normatively with nonsingular sets of independent cities, much less of interdependent cities, nor with a single city in its historical or evolutionary setting, nor with cities in terms of their relationships with the rest of the world.

Indeed, even this minimal formulation of the optimal city size concept is not very helpful, for the gap between the optimal and the actual is expressed in numbers of people. Yet clearly it must make a difference if this gap were to be filled by infants in arms, engineers, housewives, retired policemen, or junkies, or mixtures of same, not only because the change in the city's gross product would depend on the compositon of the incremental population (which must certainly be one, if not the only, element in local social welfare) but also because of their differential capacity to produce externalities and the different ways in which externalities would affect their welfare.

The issue assumes another dimension if we view the city as one region in a two-region national economy where productive factors(and, hence, population), income, interregional transportation costs, technology, and consumer preferences are fixed.

In this case "optimal" city size has been defined in two respects. The first definition answers the question: how can productive activities in this economy be allocated between the two regions so that the welfare of the city region is maximized, where welfare is objectively measured by some aggregate such as gross regional product? A formulation which can best be described as "colonial" because of its acceptance of interregional income disparities, which deprives it of the quality of optimality. The second case can be distinguished from the first by the question: how can productive activities be allocated between the two regions so that the joint welfare of both is maximized? The size and composition of the economy of the city region would not be identical in the two cases. The first case has long been a concept of interest to city planners and inperialists with all of its suboptimal implications. The second case more closely resembles the problem defined by the proponents of a national urban growth policy for the United States.

While the usefulness to policy of the conventional concept of optimum city siz is questionable, a more persuasive logic is associated with the concept of an optimal settlement pattern for a given set of economic activities and consumer preferences which would be characterized by the condition that no firm -- or worker -- could so change location as to increase total welfare. Under such conditions every settlement of whatever size is optimal, a consideration which suggests that policies seeking simply to reduce the scale of very large cities and to increase the size of settlements at the lower end of the size distribution have no special normative virtues. At a particular moment in the history of a given economy an optimal size distribution of settlements will be defined by the then characteristics of the economy, and the departures of the actual from the optimal will be a function of imperfections in the market processes. There is nothing sacred or metaphysical about the rank size rule, nor are there *a priori* grounds for concluding that the high degree of economic concentration known as primacy in some developing countries departs substantially from an optimal settlement pattern.

The formulation of the settlement pattern problem which goes along with the prescriptions for a national urban growth policy generally treats the settlement pattern as a highly interdependent system, whose principal elements are urban and rural settlements containing factor endowments and productive activities which add up to national totals and exhibiting specific relative welfare characteristics. In an optimum arrangement at the margin no worker or dollar of investment could be reassigned in a manner that would result in an increase in total welfare (here the sum of money and psychic income). Such a state of affairs would maximise not GNP but some weighted sum of GNP and other, extra market satisfactions; indeed, maximizations of GNP will result in a welfare optimum only in the special case of a tight complementary relationship between GNP and these other sources of private satisfaction.

Our growing sensitivity to the damage inflicted by such big city externalities as congestion, environmental pollution, social disorder, physical blight, noise and vibration on peoples' general levels of satisfaction justifies heavier weighting of the psychic income factors in welfare compared to objective welfare measures such as GNP. More specifically, the fact that deflated money income rises with city size in the U.S., as Irving Hoch and others have shown, could be interpreted to mean not only that large cities are more productive than small, but that firms reaping the rewards of that enhanced productivity to do so only by "bribing" labor with higher wages to leave smaller (and presumably psychically more satisfying) communities. If this properly characterizes this relationship, then two conclusions follow: (1) each worker relocates in accordance with his own preferences for combinations of money and psychic income (which may explain some part of the income differences between amenity-rich southern regions of the U.S. and the high-income Northeast); and (2) the evaluation of the net of disadvantages to a worker, working in large and inhospitable cities is bid into the cost of production of goods and services produced in such cities. Where such goods are destined for the local market these diseconomies become reflected in the cost of living.

Where they are exported to the rest-of-the-world, the purchasers of the city's products quite properly bear the costs of these externalities. There is, hence, a market mechanism which can monetize in part the nonmarket costs and benefits to urban residents of major urban externalities, and increases in net psychic disbenefits from further urban growth will show up either in upward pressure on wages or outmigration of labor to jobs in other communities, where the combination of wages and psychic income is more consistent with the subjective weighting accorded by individual preference functions.

A possibly major, if not the exclusive, path by which externalities enter into production costs, then, is through the labor market, and hence it is germane to speculate how such processes may be equilibrating and how they may affect the structure of the urban economy. In urban economy firms which are relatively insensitive to the increasing diseconomies of scale of urban areas can be distinguished from those whose production decisions are affected by changes in the levels of negative externalities experienced. The "insensitive" sector would be characterized by high capital-to-labor ratios, to the extent that the major externality impacts are transmitted through the labor market. We can further distinguish two kinds of sensitive firms: those for whom increasing negative externalities result in the upward shifting of their production cost curves either through rising labor costs or through other kinds of impact, or innocent firms, and firms whose production cost curves will shift upward as they reduce their output of negative externalities, or noxious firms. The relationship between these two sectors is that under the existing structure of property rights innocent firms are called upon to subsidize the output of the noxious sector, with the result that the local economy produces too much in the way of noxious goods (and hence too much in the way of negative externalities) and too little of innocent goods. As cities grow, then, there is some propensity for innocent firms to be squeezed out because of the secularly rising costs of production. In that subsector of innocent firms selling to local markets prices will rise and output and employment will fall; in that subsector selling to the rest of the world these costs will be passed on to consumers where price-elasticities of demand are very or will result in loss of markets and contraction of output.

Under these conditions innocent firms will tend increasingly to relocate in smaller communities with lower levels of negative externalities.

These generalizations do not clarify whether or not the existence of such externalities results in cities having too much population -- a conclusion which would depend on the relative sizes of the noxious and innocent sectors. Following Tolley's reasoning, the imposition of a tax on the noxious industries (or the sale of the right to pollute) for the purpose of compensating the innocent firms would result in contraction of the noxious sector and expansion of the innocent sector in the city; whether the net was a gain or loss of population would depend on other factors not specifically considered up to this point.

In general, then, it is not obvious that the existence of externalities results in cities of larger than optimal size so long as capital and labor are reasonably mobile. It is also admissible that externalities can result in cities of less than optimal size. It does seem clear that these negative externalities can bias the composition of the local economy in a way that may result in some specialization by city size groups.

Optimum city size, hence, only has meaning useful to national urban growth policy deliberations when it is defined as the state of subsystem (the specific city) which is consistent with the optimal state of the total system (in this case, the national settlement pattern). This rules out of the arena concepts which specify magnitudes of urban settlements as the desiderata to be pursued by such policies.

(2) Private costs and benefits as a function of city size. The rhetoric surrounding urban growth policies charges that people in large cities are worse off than they would be in smaller cities and that the costs of production are unnecessarily high because of the obvious inefficiency of such cities. The evidence that we have is certainly not consistent with this hypothesis, although it is not conclusive. To the extent that money incomes are measures of the welare of persons, Hoch, Fuchs, Alonso, and others have shown that in the U.S. deflated per capita incomes vary strongly with urban scale, once the north-south differential is accounted for.[40] In addition, recent studies of rural-urban migrants have indicated that their incomes quickly converge on those of the residents of the destination city, and hence vary with city size.[41] To the extent that these income differentials reflect increasing unit costs of labor with city size, both labor and capital must be more productive the large the city under the assumption that labor and capital are mobile. Otherwise, disinvestment would take place, unemployment would increase and/or wages decline until an equilibrium was effected such that the difference in incomes between that city and other approximated again the costs of migration. Indeed the only way to explain the continued growth of large cities is that the returns to private agencies of labor and capital are sufficient to continue to attract investment.[42] Some contrary evidence might be adduced from the fact that the very large cities in the national economy tend to increase in population at a substantially lower rate than the next tier of cities, but their total share of growth is so large as to leave little doubt about their productivity as a class.

While there is some quite good evidence about the implication of city bigness for money income to individuals, less is known about the direct effects of city size on the behavior of firms. Little is to be found in the literature about the distribution of establishments by industrial classification and by size among cities by size. Such information would be especially interesting with respect for firms producing for nonlocal markets (or classes of firms having high location quotients) since it would permit inferences about the competitive advantages of cities of various sizes for these kinds of industries and suggest how growth is likely to affect industrial composition. Such analysis would be a step toward integration of size-of-city variables with models of interregional trade, industrial location, and urban hierarchies, and so increase the analytical power of models of the urban system.

(3) Externalities and city size. Overstatement of the national welfare costs of those urban externalities which vary with city size is endemic in much of the casual analysis of the consequences of city size. The growing interest in the analysis of how technical externalities vary with city size has largely addressed negative externalities, while positive externalities are usually dismissed as agglommeration economies or urban amenities. While this disparity of interest can be explained by the policy issues surrounding congestion, environmental pollution, reduction of longevity, and negative social behavior, the fact remains that large cities have exerted an attraction to persons and enterprises which cannot always be explained by private scale economies. The "bright lights", the stimulation, the vast array of opportunities to indulge even the most curious interest or specialized talent give the environment of large cities the quality of pure public goods which appear to be highly valued by many: the rich configuration of services, the reliability of the labor market, and access to other closely linked activities act in much the same way for firms. The importance of this rather commonplace observation is that the emphasis on negative externalities in the appraisal of welfare aspects and behavioral features of urban scale leads to a partial analysis which overstates the the costs of urban scale.

While money costs and returns can be expressed in net terms, we have no direct parameters for offsetting positive and negative technical externalities against each other. To the extent that they are felt in the costs of production, they become embodied in the product and objectified in the market thus being transformed into pecuniary externalities which can either be exported or enter into the local cost of living. To the extent that they affect the utilities of the individual, they become netted out by the way in which he trades off net money income, negative externalities, and positive externalities of urban scale in making economic choices. Hoch's analysis mentioned earlier, indicates that externalities associated with urban scale appear to net out negatively in the labor market, the higher real income earned in larger cities can be interpreted as including compensation to the worker for loss of utility he experiences in the large city.[43] Thus, the labor market provides a mechanism for monetizing urban externalities. Within cities transactions in the land market also monetize local differentials in positive and negative externalities.[44] Indeed, the diffusion of the impact of urban externalities through market processes makesit enormously difficult to pursue microeconomically their incidence and, hence, to quantify an "actual" income figure which has welfare comparability among cities in an urban system. Tolley has described the amplifying effect of local externalities embodied in goods for the local market, which suggests that cost of living and the real income increments of urban scale are closely related:[45] if the externality effects netted out to zero in City A, any scale-associated cost of living differential would have to result from rising cost functions in the production of local services.

All of this suggest that if local export industry is competing effectively with that of other cities, it should be possible to quantify roughly the net social costs of city bigness by assuming that wage differentials by city size maintain a common welfare level at the margin among cities. The sum of the differentials over the intercept then might be interpreted to be a measure of urban scale diseconomies. However, while the marginal worker may be indifferent among combinations of city-size diseconomies and wage differentials, all workers are not necessarily so characterized. In a particular city workers for whm the disutilites of negative urban externalities are lower (or for whom the utilities of the city's positive externalities are greater) will enjoy a surplus which can not properly be charged to the social costs of urban scale. The amount of this surplus is measured by the amount wages would have to be reduced before a worker would migrate.

Nevertheless, city-size income differentials make possible some first approximations to the external costs of urban scale by definition of an externality-free, basic income per capita. Such a formulation allows us to specify alternative arrays of city sizes and industrial compositions in a manner conducive to the evaluation of incremental ways of improving the efficiency of the settlement pattern as a whole.

Behind the exercise lies another interesting question, what determines the slope of the city size cost function? Transportation costs will appear among the variables to the extent that they affect the costs of living among cities. This is not to say that transport costs are necessarily associated with urban scale but rather they they reflect generally the spatial organization of the economy: lower transport costs should be manifest in a lower slope in the city-size cost function. In similar fashion, migration costs should find their way into the shape of this function, since they limit at the margin the mobility of the marginal worker. Industrial composition by city size will affect the slope, but it will include some monetization of externalities, and it seems not unreasonable to think that this slope does afford a relative measure of the next diseconomies of urban scale. If there are substantial diseconomies to urban scale, one would expect this curve to approach some minimum slope dictated by migration costs at the upper end.

It is further worth distinguishing what our environmental economists are coming to call "emission functions" and "damage functions," the former expressing the pollution-generating characteristics of a set of activites, the latter defining the welfare impacts of volumes of pollutants on human populations, given intervening variables, such as local climatology, population characteristics, and the like. Similarly, the welfare impacts of the externalities associated with urban scale are composed in general of externality production functions and consumption functions, and social costs are probably not simple monotonic functions of the volume of physical externalities, for the simple reason that individuals and firms will effect tradeoffs to minimize damage imposed by externalities. Thus in the longer run while the time costs of congestion tend to rise with increases in urban scale, and while the marginal value of labor function suggests that up to a point each successive unit of time lost in transportation is more valuable than its predecessor, workers can compensate in part by carpooling or using public transportation, that is by trading off among vehicle operating costs, the utility of driving ones own vehicle, and time costs. The resulting adjustment must, by definition, be superior in welfare terms to the preceding state. Thus, the assessment of externality costs proportional to the volume of the externality results in a further overstatement of the social costs of urban scale. This observation makes it important to know more about externality generation functions and related damage functions not only to understand behavior more precisely, but also to clarify the welfare consequences of urban scale differentials.

(3) Consequences of anti-externality policies While it is possible in the abstract to speculate about the characteristics of a city system in a perfectly competitive world, our interest in policy makes us begin with the historical given of the current situation, and it is not obvious how recitification of the market imperfections in the real world would affect characteristics of the urban system. Would more or fewer people live in the metropolitan sector? How would the city-size cost function be altered? How would the rank-size function of cities be modified? How would the internal organization of urban areas be modified? While a more efficient allocation of resources would

be achieved by internalizing the external costs of urban activities, it is not always obvious what the longrun equilibrium would look like. In the analysis discussed at an earlier point. Tolley deduced that internalization of externalities would raise marginal cost curves, reduce outputs, and result in a downward shift in the size of larger cities as resources were reallocated to activities in smaller cities with lower levels of external costs; it follows that abatement of externalities would result in larger cities becoming even larger. (and smaller cities smaller, presumably). Actually a public policy of regulating the production of externalities would save some mix of these effects, and the consequences for city size would depend on a number of other characteristics, such as how the policies affected polluting vis-a-vis nonpolluting activities, industry mix, and the extent to which such policies were applied to one city or to the entire system.

The nature of property rights defines the incidence these effects. Common property aspects of the environment have made it possible for polluting industries to shift costs to the community at large, and the institutional definition of some public services as collective goods has provided another mechanism for diffusing private costs through the community as a whole: the collective good characteristics of public streets encourages congestion costs and diffuses them widely throughout the community. Internalization policies would involve substantial changes in property institutions and so have broader consequences for the settlement pattern than those effected directly by the abatement or internalization of external costs. As public interest grows in such institutional changes, it will be increasingly important to keep in focus the tradeoffs between the gains from controlling externalities and the inadvertent losses in terms of the attainment of other societal values.

A word about the distributive consequences of externalities. If the hypothesis can be validated that external costs are distributed in a manner inverse to income, a number of consequences follow. The proposition itself is not unreasonable, since one of the goods that can be bought with money in this society is an environment relatively free of external effects -- the more income you have, the greater chance you have to command "life activity" locations in the urban space which reduce your exposure to external costs. Hence, the structure of rents is influenced by the spatial distribution of the external costs which bar the poor from these areas (for the rich the poor themselves may be a negative externality). The reduction of urban externalities can be expected to benefit the poor in relative terms and so advance the reduction of class disparities in the large cities. Income redistribution policies, on the other hand, would allow the present poor to compete more effectively with the present rich for the less externality-ridden locations in the city and so drive up the rents, reward rentiers for their foresight, and result in more intensive development of such areas. In short, intraregional consequences of urban externalities and policies to internalize them warrant further investigation.

Analysis of urban externalities has special implications for the New City case. Overall design, the control of activity mix, and the political ability to make the quality of the environment a pure public good subject to public control might very well reduce levels of scale-associated externalities substantially, with the result that the efficient scale of such cities might be very large in the absence of any effective counter-externality policies in the rest of the urban system. Hence, the appeal of current thinking about optimum city size may lead to substantial under-design of such cities. It is clear that the real constraint on the size of planned new cities is financial: front end costs for large cities appear to go far beyond what private sources can mobilize and sustain over long repayment periods. Nevertheless, the inference is hard to avoid that national policy directed toward the development of new towns and small cities would be misplaced, and the national government ought to be more concerned with exploiting the twin efficiencies of large scale and planned development.

(4) City-size distributions. This paper has touched on the analytical dimensions of the size distribution of cities in a national system. The dominant concepts here are the hierarchy of cities and the rank-size relation, which have been forumlated by geographers and other to describe rather than to analyze the urban system. Lasuén has presented evidence that national urban systems measured in terms of rank-size relationships and economic specialization are very stable over time. Nevertheless, there is precious little to be found which integrates such resource allocation processes as industrial location and internal migration to an outcome which can be described in rank-size or urban hierarchy terms. Clearly anything that throws light on the determinants of urban scale will enrich understanding about the distribution of city sizes, and it is this understanding which is an essential element population distribution policy.

(5) The costs of public services and the size of cities. We have not yet developed a term to describe the financial system which is made up of many thousand local jurisdictions, who are the basic producers of local public services, and their financial relationships with state and local government. A model of the system would have the following general form; the supply side being expressed as

$$a_{ij} = f(b_{ij})$$

$$b_{ij} = g[e_{ik}, (s + n)_{ij}]$$

where a_{ij} is a measure of the level of output of public service j in local jurisidiction i;

B_{ij} is the input of resources into function j in jurisdiction i;

e_{ik} is the number of people in i having socioeconomic characteristic k;

s and n represent resource inputs from state and federal sources,

and on the demand side as

$$a_{ij} = h(A_{ij}, Q_{ij})$$

$$Q_{ij} = p(P_i, e_{ik})$$

where A_{ij} is a measure of capacity for producing public service j in jurisdiction i;

Q_{ij} quantifies the clientele of j in i; and

P_i is the population of i.

In addition, f designates a cost function; g expresses resource availability and community preferences for service j; h represents the effects of queueing and other congestion effects on quality of the service, and p relates the clientele of service j to the size and composition of the local population. i is structured to rank localities by size and j is some very disaggregated listing of public services.

From the standpoint of national urban growth policy we would be interested in defining the structure of the relationship between a and n to clarify the consequences of different federal (and state) financial policies with respect to the production of urban overhead services. The output variable a_{ij} has a double significance. On the one hand, the objectives of federal urban policies can be expressed in part in terms of a_{ij}; certainly this is true of the categorical grant programs which were largely addressed to reducing deviations of a_{ij} with respect to particular j's. Advocates of the President's revenue sharing program argue by implication that on the whole the structure of a_{ij} ought to represent community consumption functions rather than standards imposed at the national level.

It is in the second significance that the national growth policy advocates would root their program, for the principal policy tool available to the national government with which to implement a growth policy is financial assistance from the fiscal resources available to the federal government structured to bring about a more desirable set of values for a_{ij}, for a_{ij} exerts an influence on the distribution of profitabilities and utilities in the settlement pattern. These effects may arise from direct consumption of these services; equally important, they may arise from the effect of such changes in service levels in the reduction (or monetization) of negative externalities associated with urban scale. Thus an urban transportation project may provide a different kind of transportation service, or it may reduce congestion costs in the system. To the extent that this project is subsidized by the federal government, the nation as a whole may indeed be subsidizing excessive scale in the urban system, for such a policy has increased the divergence of public and private costs and benefits. If federal policy affected only that city, the case would be comparatively clear; where national policy is so pervasive, it is implicit that the entire system is nonoptimal with respect scale but the direction of the optimum is not so obvious.

Hence, federal policy addressed to helping the cities meet their demands and needs for public services cannot help but distort key welfare characteristics of the settlement pattern. But how one normatively evaluates the outcome will depend in part at least on how one specifies the interface between national and local interests. The proliferating categorical grant program authored by the democrats saw a strong national interest in reducing the service disparities among settlements in the urban system, an interest more often than not expressed in equity terms. The current administration restricts the national interest to making our federal system of government a more effective agent for the representation and satisfaction of community wants. It is a matter for collective decision making to decide which approach to collective decision making is in the interest of the whole. Nevertheless, it is clear that alternative approaches have unique consequences for the settlement pattern. Whether they are nontrivial or not is another research question. In the same way it follows that the structure of local revenues and expenditures also has implications for the settlement pattern, not only on a national scale, but on the organization of the metropolitan regions. As things now stand, we can have little confidence in the charge that federal policies addressed to the subnational policy economy weighted the scales against nonmetropolitan America and in favor of the major metropolitan centers.

While the constraints of this paper preclude any detailed investigation of this problem, it is worth observing that there are characteristics of current policy whose consequences for asserted objectives are probably negative. Indeed, the matching mechanisms may be one of them. It seems likely that the current policy does by indirection favor the larger cities, not only because they are more capable of availing themselves of federal categorical assistance, but also because the matching requirements are more easily met than in poorer or smaller cities. An important innovation of the Appalachian Regional Development Program is the development of a fund with which the Commission can help its growth center localities meet the matching requirements of federal grant programs. One is left to speculate on the urban system impact of other such features on the subnational public economy.

(6) Migration and settlement pattern. More than one of the advocates of a national urban growth policy have proposed that there be explicit federal policies to influence the pattern of migration among the elements of the national settlement pattern. Internal migration floods New York City with welfare claimants, strips the rural areas of their prime human resources, results in a flocking of

migrants to the largest metropolitan areas imposing social costs on the older residents of these areas, an argument which seems to suggest that everyone would be better off if no one moved, and, indeed, legislation has been offered in Congress more than once recently to subsidize people not to move. More sophisticated proposals have suggested that better organization of the labor market and a system of limited inducements could redirect migrants to job opportunities which carry with them lower social costs than those resulting from migration to the major cities. These policies would seek to influence the migrant's choice of destination rather than his decision to migrate. All make assumptions about the economic irrationality of migration decisions which can stand little searching scrutiny.

The basic conceptual model of internal migration in the U.S., developed and empirically implemented by Lowry finds outmigration to be a function of age and socioeconomic characteristics of local population and essentially independent of local economic opportunites.[46] Inmigration, on the other hand is a function of employment opportunity and hence of the interareal flows of new productive investment, and Morrison has suggested that some national controls on the location of new investment could effectively alter the structure of internal migration without much in the way of explicity migration policy.[47] Implicit in these formulations are assumptions that on the whole migration decisions by individuals turn out to be inherently rational, conclusions which studies of migrants have supported.[48] Add to this the high degree of stability of internal migration rates in the United States at around six and a half percent per year and the power with which internal migration can change the settlement pattern in comparatively short periods of time becomes obvious. Twelve million people find themselves living in a different county or state than they lived in the preceding year. If no one ever moved more than once 120,000,000 people would have migrated between 1960 and 1970, yet the metropolitan regions of the United States grew in population only 15,000,000 a little faster than the nation as a whole. Subtracting out growth by natural increase, actual net migration to the metropolitan sector must have been very small, not much more than one percent.

Hence, the great bulk of these migrations must have taken place within the metropolitan sector and the ratio of gross to net movement must have been very high. This volume of movement suggests the possibility that changes in the settlement pattern could be effected relatively rapidly _if_ the destination choice could be influenced. The stability of outmigration and overall migration rates, however qualify such a conclusion, as does the prospect for any early and substantial public policy to influence the economic location decisions of the private sector, since most new job formation is taking place in the highly decentralized, locally-oriented services sector of the economy. Nevertheless, it would be useful to enrich such speculations with some empirical analysis.

The very large volume of internal migration also raises questions about the differential impact of migration processes on communities in the national system. The movement of a family from City A to City B will have the effect of shifting fiscal resources from the one to the other as well shifting the demand for public services -- where fiscal contribution by the family exceed the public cost of servicing it there is a transfer favoring the destination City B; where servicing costs exceed the fiscal contribution, the transfer favors the city of origin, City A. If there are substantial diseconomies of urban scale, migration from a smaller city to a very large city would result in welfare losses at both ends, since the small city looses a scale advantage and the net diseconomies of the large city are increased. In short, migration can reshuffle welfare through the transfer of fiscal resources and through its effects on the negative externalities associated with urban scale. To the extent that their movements are generally in one direction, the relative growth advantages of cities is being reshuffled.

To Conclude

It doen't seem very helpful to offer that the base of solid analysis on which decisions about national urban growth policy will have to rest is hardly strong enough to bear the weight. There is a growing political force to develop some overarching urban policy like postwar agricultural policy, which could make the entire subnational economy perform better. Fortunately, such a policy is not an immediate threat, which makes it possible for the state of knowledge to be brought up a bit closer to the intellectual demand of public policy. At the moment, it would be difficult to do a responsible job of estimating the impact of the categorical grant-based policy initiated under the Kennedy administration and still the basic policy of the country. It is equally difficult to test the consequences of the New Federalism for the health and welfare of the thousands of local jurisdictions that make up the subnational economy individually and collectively.

For the kind of national urban growth policy which would seek to alter the configuration of urban sizes, the pattern of development of metropolitan regions, and the comparative advantage of nonmetropolitan America in the national economy, it becomes necessary to know much more about the processes at work influencing the size of settlements and the size distribution of those settlements, as well as about the system. The production and welfare impacts of urban, scale-associated externalities as well as those of declining scale which typify nonmetropolitan America are not entirely a dark intellectual continent, but there is much to be done before policy models can reflect such features with any degree of reliability. The processes which shape the settlement pattern -- public and private investment and internal migration need a higher degree of predictability before the consequences of policy intervention can be anticipated.

Subjecting the national settlement pattern to public policy will require a flow of research and model building similar in character to and substantially greater in scale than that which resulted from Bureau of Public Road's discovery of the land use-transportation model. Properly institutionalized, such an effort could help shape the content, as well as objective function, of national urban growth policy.

June 2, 1971

[1]Public Law 91-609, 91st Congress, H.R. 19436, December 31, 1970

[2]Ibid., Section 703

[3]Message from the President of the United States to the Senate on Special Revenue Sharing for Urban Community Development and Planning. Congressional Record, 92nd Congress, First Session, Vol. 117, No. 29 (March 5, 1971) S2503-S2507. The "number of steps" apparently refers also to the content of his message on Special Revenue Sharing for Rural Community Development, which was submitted a few days later.

[4]Ibid, p. S2505

[5]Lloyd Rodwin, Nations and Cities: A Comparison of Strategies for Urban Growth, (Boston: Houghton Mifflin Co., 1970)

[6]The thirty million people on farms in 1940 would have grown to sixty million under conservative natural increase assumptions and zero outmigration rates. The present rural farm population is less than ten million. Even so, Irving Hoch argues that farm price support policy inhibited the transfer of labor from agriculture to nonagricultural sectors.

[7]Report of the National Commission on Civil Disorders (New York: Bantam Books, 1968)

[8]Donald Canty, "What Is This Thing Called Urban Growth Policy?" in City, August-September 1970, 31-32

[9]Kermit Gordon (ed), Agenda for the Nation, (Washington, The Brookings Institution, 1968)

[10]The National Commission on Urban Problems ("Douglas Commission"), Building the American City, House Document No. 91-34, 91st Congress, 1st Session (Washington: USGPO, 1968)

[11]The President's Committee on Urban Housing (Kaiser Committee), A Decent Home, (Washington: USGPO, 1969)

[12]U. S. Department of Agriculture, Communities of Tomorrow, Agriculture/2000 November 1968 (Washington: USGPO, 1968)

[13]National Growth and Its Distribution, Proceedings of a Symposium on Communities of Tomorrow jointly sponsored by the United States Departments of Agriculture, Housing and Urban Development, Health Education and Welfare, Commerce, Labor, and Transportation, in Washington, D. C., December 11 and 12, 1967.

[14]Ibid., pp. 4-6

[15]Advisory Commission on Intergovernmental Relations, Urban and Rural America: Policies for Future Growth (Washington: USGPO, April 1968)

[16]Ibid., pp. 129-130

[17]U. S. Senate, Committee on Government Operation, 90th Congress, Second Session, The Rural to Urban Population Shift: A National Problem;Proceedings of the National Manpower Conference sponsored by the Senate Committee on Government Research, the Ford Foundation, Oklahoma State University and held at the Student Union, Oklahoma State University, Stillwater, Okla. May 17-18, 1968

[18]The President's Remarks at Ceremonies Marking the 100th Anniversary of Dallastown, September 3, 1966, Weekly Compilation of Presidential Documents, September 12, 1966, p. 1217

[19]Donald Canty (ed), The New City (New York: Praeger, 1969)

[20]This proposal gained an important adherent on February 16, 1971, when David Rockefeller publicly accepted the goal of 110 New Cities and made specific organization proposals for implementing the program. New York Times, February 17, 1971

[21]As reported in City Chronicle: Monthly Report of Urban America May 1969, pp. 1-5; also, revised and published as

Daniel P. Moynihan, "Toward a National Urban Policy" in Public Interest, Fall 1969, pp. 3-20; also published in Daniel P. Moynihan (ed), Toward a National Urban Policy (New York: Basic Books, 1970) 3-25

[22]Ibid., p. 4

[23] As reported in New York Times, February , 1970

[24]Congressional Record, 92nd Congress, First Session, Vol. 117, No. 29 March 5, 1971, pp. S2503-07. In the White House briefing Secretary Romney described the present system as resembling "a 20-mule team harnessed in the dark by a one-eyed idiot." New York Times, March 5, 1971.

[25]The recommendations in this message were modified a few days later in the President's message to Congress on Special Revenue Sharing for Rural Community Development in rectifying an apparent oversight. In the later message the President recommended the addition of $100 million to go to small cities in the 20,000 to 50,000 range who have been receiving funds under HUD programs but would not qualify under the special revenue sharing for Community Development and Planning.

[26]In addition, the message anticipates legislative proposals to support the preparation of statewide development plans as qualifying documents for these special revenue sharing program. Cf. Congressional Record, May 10, 1971, H1375-8

[27]President's National Goals Research Staff, Toward Balanced Growth: Quantity with Quality (USGPO, Washington, July 4, 1970)

[28]Ibid., p. 54

[29]Ibid., p. 58

[42]Recent net outmigration of firms and loss of 72,000 jobs from New York City are clearest evidence to date that New York has gone beyond its optimum size. cf. New York Times, May 1971

[43]Hoch, op cit

[44]Ronald G. Ridker, Economic Costs of Air Pollution, (New York: Praeger, 1967) Chapter VII; Kenneth F. Wieand, "Property Values and the Demand For Clean Air: Cross Section Study for St. Louis" A paper presented at the Committee on Urban Economics Research Conference, September 11-12, 1970, Chicago, Illinois; and John A. Jabsch, "Air Pollution: Its Effects on Residential Property Values in Toledo, Oregon" The Annals of Regional Science, Vol. IV, No. 2 (Dec. 1970) 53-67

[45]George S. Tolley, "The Welfare Economics of City Bigness" Urban Economists Report #31, University of Chicago, December 1969

[46]Ira S. Lowry, Migration and Metropolitan Growth: Two Analytical Models (San Francisco: Chandler Publishing Co. 1966)

[47]U.S. President, Special Revenue Sharing Program of Rural Community Development, Message from the President of the United States to the Congress, as reported in the Congressional Record for March 10, 1971 pp. H1375-8 and S2867-70

[48]Cf. esp. Richard Wertheimer, The Monetary Rewards of Migration Within the U.S. (Washington: The Urban Institute, 1970)

June 2, 1971

27 May 1971

The National System of Cities as an Object of Public Policy

Wilbur R. Thompson *

NOTE: This paper has been prepared for the Resources for the Future-University of Glasgow Conference on Economic Research Relevant to National Urban Development Strategies, Glasgow, Scotland, August 30-September 3, 1971. It is subject to revision and should not be cited or quoted without express permission of the author.

* Professor of Economics, Wayne State University, Detroit, Michigan

Wilbur Thompson

Born in Detroit, Michigan in 1923 and a resident of that area with only occasional interruptions ever since, until taking up (semi-permanent) residence in Phoenix, Arixona in 1970. A graduate of Wayne University in Detroit (B.A., 1947) and the University of Michigan (Ph.D., 1953) and a member of the economics faculty of Wayne State University since 1949. A research associate (on leave) at RFF in 1961-62 and 1964-65; Chairman of the Board of the Southeastern Michigan Transportation Authority (1968-70); and currently lecturer and consultant to the Urban Journalism Program of Northwestern University and The Urban Policy Program of the Brookings Institution. Author of _An Econometric Model of Postwar State Industrial Development_ (1959) and _A Preface to Urban Economics_ (1965) and various articles on urban and regional economics.

Many are impatient today with questions such as: how big should cities be, what mixtures of work should they perform, and where should they be located? This emphasis sounds too dated and too closely related to the impersonal focus of the earlier city planner who was almost wholly a physical planner. It is people not places that demand our attention, the new breed of urbanists sternly insist. It is poverty, race, crime and the quality of the environment that are at issue.

This charge can not be answered without first admitting that at many times and in many contexts -- in policy contest with local merchants, landlords, land-owners and public officials -- vested business, property or job interests of people and places are in sharp conflict. But in more extended discussion among persons with less vested interest in particular places (academics, with little at stake any place?), it should be possible to reaffirm quickly the primacy of people and then to move quickly on to a consideration of the size, function and location of cities as a means to good ends.

We might choose to save some small place because we wish to save some elderly poor person an expensive and unnecessary move, or we may wish to conserve a stock of good, used housing during a housing shortage. Or we may elect to shut down a small place to break a cycle of poverty that threatens persons yet unborn. We may limit the size of a city or empty out a fragile natural area to avoid creating an ecological "hot spot" that would be very costly or impossible to reverse. We may see in a national settlement policy the welfare of the next generation. It is less a question of what is relevant to people and more one of the length of the planning period; land planning in a national context has a very long pay-off period, greater than even educational investments in most cases. It is in this spirit that this paper is written.

27 May 1971

City Size and Consumer Welfare:
Variety versus Economy

From the vantage point of the household, the gains from larger city size probably vary with socio-economic status. Scale brings variety for greater consumer choice, and sheer numbers of sellers promote competition in price and quality. (The small town provides the best illustrations of "local monopoly" outside of the regulated public utilities -- what small town has a "discount house"?) But the advantages of variety vary directly with education and income; it is the more affluent households who in general have most to gain from this attribute of large city size. In sharp contrast, the lower income household achieves economies in the purchase of standardized goods and services ("necessities") with increased size of the local market, but the favorable effects of competition on consumer welfare probably flatten out rapidly at moderate city size (100,000 population?). Further, the advantages of competition and economy are inversely related to income and diseconomies of scale may begin to set in at a population of about one million, especially for the ghettoized poor. The deterioration of public transit often confines the poor to a sub-market of the great city and subjects them to big city counterparts to the local monopolies of the small town, made more pernicious by the impersonal and more exploitive character of unresponsive, absentee ownership.

Economy in retail trade and personal and professional services may then be "U" shaped, in sharp contrast with variety which in general increases, with scale, virtually without limit. To these more objective variations in consumer welfare with city size, we must, of course, add the more subjective variations in life style that go along with different city size. Even so, the platitude that we need a wide range of city sizes in which to express a wide variety of life styles seems to miss the main public policy issue. The normal market process has given us this. The problem is, apparently, that the number of lower income and/or less-educated households that prefer or are indifferent to large city living is less than the number of such households that are "needed" there -- industrially and occupationally linked to the more-educated, higher-income households who prefer large cities. Conversely, too few of the high status group are willing to live in smaller places and generate work for those less able and therefore locationally dependent. Consequently, many less-educated, lower-income households are "forced" to live in the large cities that the higher status group prefers.

The conflict is resolved in favor of the educated, affluent group, in part because they supply the relatively scarce professional and technical labor and have therefore greater leverage in directing industrial location. But perhaps just as important, the great mass of semi-skilled production workers have implicitly given up whatever influence they might have had on the location of their work (and residence) by quoting a spatially invariant wage through their unions. What is the effective meaning of the Gallup Poll finding that 56 per cent of the respondents would prefer to live in a small town, when they say, in effect, through their supply price of labor, that they do not care where they live?

If it is unlikely that organized labor would be willing to experiment with geographical wage differentials sufficient to induce the relocation of industry, the national public policy issue would seem to be whether alternative ways might be found to register labor cost differentials that would guide production into locational patterns that would better reflect household living preferences (i.e., raise real wages). How much money wage (after taxes) would various groups of workers trade for the objective and subjective gains from living in smaller places, if a stable way of accomplishing this could be found -- cut-throat competition in wages prevented?

Urban Economic Scale and Local Government

We have had a number of inconclusive studies on the relationship between city size and public expenditures, and by implication the efficiency of the local public sector. Those local public services that are like, or are in fact, public utilities (e.g. water, sewage disposal and public transportation) do show economies of scale up to populations of medium size.[1] But the cost functions of the more critical public services (e.g. education and public safety) are much more elusive. Besides, it is more the effectiveness (quality) than the efficiency (cost) of local "government" that is at issue. What can we say about how "good" local government is at various city sizes?

Surely, the skill and sophisitcation of local public management tends to increase with city size, but then so does the complexity and difficulty of the work. And surely it is not only the lure of higher salaries but also that very challenge of complexity that serves to attract the most able public managers to the bigger cities. But the net balance of the "supply and demand" for skill is quite unclear. This is due in part to the fact that the strength bargaining power -- of the local public sector, vis-a-vis the private business sector, also varies with city size. A priori, one might argue that small places are especially vulnerable to the footloose manufacturing firm that threatens flight if, for example, taxes are raised or pollution standards enforced. And, at the other end of the size spectrum, local governments in large, politically-fragmented metropolitan areas feel no freer to pursue tax equity or to guide location and land-use with a firm hand because they fear the loss of tax base through flight as keenly as remote small towns fear job loss. (More often than not, manufacturing firms do not yield a net fiscal gain to small towns because of tax-exemptions or below cost rentals of community owned plants employed to attract them; this is much less true of political subdivisions in large metropolitan areas.)

This all seems to suggest that middle-size urban areas of a couple of hundred thousand population, on up to perhaps a million or so, have the advantage of remaining relatively whole governmentally while becoming moderately stable economically. This would seem to provide the strongest and fairest hand in taxation and public regulation. Returning to the prior consideration of attracting high talent into local public administration, "countervailing power" should count heavily.

One does not need to resolve the relative importance of cost-efficiency versus policy-effectiveness to argue that the latter is more central to a consideration of national urban policy. A responsible Federal government should care most of all about the ability of local government to reinforce national policy, or at the very least the awareness to not counteract it. We can not re-make the country into a monolithic system of middle-size cities, even if we should so desire, but the Federal government can act to strengthen the hand of local government at the ends of the size spectrum. The rebirth of metropolitan area government in the form of "two-level local government" is a step in that direction, presumably a lighter and more measured tread.[2]

Changing Criteria in Local Labor Markets

Larger city size improves local labor markets in important ways, although some very new dimensions of scale may come to dominate. Remote, small places have always borne the burden of monopsony in the local labor market. We still have "company towns," especially in early-stage processing industries that stay close to the basic raw material, but an ever more pervasive, nation-wide unionism is equalizing bargaining power in these places. A monopoly-monopsony stand-off in price power does, however, nothing to reduce the employment instability of a one-industry, one-company, one-plant town which runs the triple risk of a fading product or an incompetent management or an obsolete facility.

Today, it is more the variety and quality of local employment opportunities in the smaller urban area that creates cause for concern. The narrow range of occupations that emanate from a narrow product base offers little opportunity for up-grading on the job, a still important alternative to formal education. But probably most important of all, especially in days to come, is the placement problem posed by the highly educated husband and wife team -- professional or technical labor in joint supply. If females would resign themselves to roles as teachers and nurses, if they insist on becoming educated and professional, joint placement would be easier. There are schools everywhere and a clinic, if not a hospital, almost everywhere.

But reflect on the vocational problem of the psychologist wife of the chemical engineer whose employer selects a small town plant location (near the basic raw material). The probability of both positions appearing in the same urban place is the product of the two probabilities for that size place. But an even greater scale is required to generate enough positions in each occupation so that there is a high probability that both positions will be open simultaneously, or nearly so. Again, while the chemical company may make provision for continuing graduate education in chemical engineering, what are the prospects for post-degree work in psychology in a remote small town?

27 May 1971

On the Outlook for Remote Small Places

The Three Fates

In the absence of decisive public policy, the typical remote small town appears to face one of three fates: de-population, socio-economic deterioration or economic absorption. Towns in very sparsely populated regions will ordinarily experience out-migration to larger places much in excess of the slow inflow from nearby smaller places. While an out-pouring of local residents can be somewhat slowed, a substantial gross out-migration is inevitable, as simple reflections of occupational and life style preferences and just plain restlessness. With few "lower" places to draw from and virtually no basis for attracting population from other similar size places, de-population is inevitable and inexorable.

A second set of towns in still densely populated rural areas, especially those with high rates of natural increase (e.g., the Deep South), will tend to experience gross inflows that approximately balance the outflows to larger places. The problem posed here is that, typically: (1) the educational level of the town is higher than that of the countryside, and (2) those who stay on the farm tend to be the more educated and skilled part of the rural population (i.e., the better farmers), and (3) those who move from the town to the city come more than proportionately from the upper half of the high school graduating class. Thus the town trades the top layers of a superior educational stock for the bottom layers of an inferior one. Towns in this context remain "alive" by being transformed into low-skill, low-income, isolated ghettoes -- under-development traps.

Many "remote" small towns are nearly ripe enough to fall into the expanding commuting range of the nearest metropolitan area; they will be rescued by transportation improvements before they have time to de-populate to a point of no return. Such a town will tend to become a specialized part of a more complex system -- typically a dormitory satellite.

Euthanasia

The appropriate public policy responses to recent developments in remote small places is much too complex to be detailed in so short a paper and too much beyond the present state of knowledge to be discussed definitively in any length paper. Still, the three fates sketched above suggest the broad outlines of three quite different policies and strategies.

No one should still be surprised by the pervasiveness of absolute decline in population; no statistic has been more quoted over the past few years than the fact that one-half of our counties lost population during the 1950-60 period. Nor will the almost inexorable power of this trend be obscured when the 1970 data have been assessed and popularized; two-fifths of our counties have lost population for three successive decades. Certainly, an official de-population policy and strategy would be political dynamite, but why is there virtually no literature on the graceful abandonment of obsolete places, created by those who are not in a politically sensitive position.

In general, the current position of students of regional development is that capital investments in semi-permanent infrastructure -- interpreted broadly to range from transportation and utility systems to technical institutes and housing -- are most appropriate for small cities that have been designated as "growth poles." Investments in health and education, especially in younger persons, increase occupational and geographical mobility and are therefore most appropriate for remote small places with bleak futures, as well as the more direct and obvious relocation allowances.[3] As general policy, so far so good; but it is an operational strategy that is so needed and so unclear.

A few illustrations will, however, suggest the kinds of questions that we need to address seriously. At what point in the depopulation of a settlement is it appropriate for the next larger unit of government to assume responsibility for the financing and/or the provision of a given public service? Just as the county sheriff was responsible for the safety of the rural population before incorporation, so too the sheriff could resume that function with contraction. Should the state reserve the right -- practice the policy -- of revoking local government charters at some given stage of contraction?

Turning from institutional arrangements to human resources, the order-of-march in migration, by age and education and whatever, certainly affects the stability of the process. Clearly, the normal sequence of migration in which the younger and the more educated tend to leave first is one which threatens a progressive deterioration -- a cumulative disequilibrium -- and virtually ensures a "premature" collapse, defined here as an unnecessarily accelerated obsolescence of capital. Could some significant modification of the normal sequence of out-migration be achieved, at reasonable cost? Would it be efficient to pay premium salaries to a few key talents (teachers, counsellors, doctors ?) to induce them to stay on longer (or come) in the contracting community, to preside over the orderly amortization of sunk capital and of immobile people?

While we may want to slow the exodus of certain talents, we may want to speed the migration of many others. What is the appropriate rate of contraction under various circumstances? De-population might be speeded appreciably if the transfer of home ownership under favorable terms were made easier. This is especially so in times of housing shortages and credit stringency when a better job may be offset with a much poorer house in the new place. A liberal national housing policy would seem to be a necessary concommitant of a tougher national settlement policy. True, we do not know much about the euthanasia of obsolete settlements, but it seems that we do not even want to talk about it.

Small Towns as Half-Way Houses

The policy-problem in the case of remote small places in overpopulated rural areas is that the locality is beggared in the act of serving the nation as a half-way house between the fields and the factories. The paradox of migration is that everyone could become better-off -- the rural migrant moving in, the out-migrant to the city and even those who stay and rise in the local hierarchy, due in part to the out-migration of the strong -- the nation, as a simple collection of these parts, would be richer because the allocation of labor would be better, but the locality as a place could be worse-off. Those who move into a community do not have to be less educated or poorer than those they join to impoverish a community, merely of lower socio-economic status than those they replace -- the top of the stock that moved out. We do not have to put places ahead of people to recognize that places are environments for people, who may thrive or languish on the basis of rich or poor personal contacts and public sectors.

We are probably fortunate, as a nation, that localities can not prevent the in-migration that beggars them, but we err in not providing compensatory payments to these staging areas, not just for the sake of equity in income redistribution but even more to provide incentives to localities to undertake the critical functions of education and acculturation less reluctantly and more effectively. The nation should monitor the gross flows of migration at the local level to know better which of its many agents in human resource development are most strategically located: typically and paradoxically, those communities most impoverished in interregional trades in human capital.

Can we express national development goals in these localities in income terms? Surely, we are not trying to maximize per capita income in any given area because that criteria could be perverted into a beggar-thy-neighbor strategy -- skimming-off the cream. We are not even trying to maximize the income of the current residents of the area because these are open economies with changing populations. The most general goal that could be defended as in both the local and national interest is to create an environment that will produce the greatest learning experience for all who pass through or stay. More operationally, the strategy might be to ignore average income relative to the nation and to concentrate on moving people upward in skill and outward in direction, if this place has the primary function of first emptying out its hinterland (migration-shed) and then emptying itself out when that job is done.

But if this place is programmed to be a permanent settlement, then converging on the national average income would seem to be a legitimate test. The corollary to that proposition would seem then to be that we should not accept, as inevitable and permanent, places that are probably never going to be able to attain an average per capita income. Or is this position too extreme? Perhaps not, if we normalize per capita income for the local industry and occupation mix and correct for the local cost-of-living, including amenities.

From Small Town to Satellite

Being rescued from oblivion by a transportation improvement

is not an unmixed blessing for the (formerly) remote small place. That settlement will almost certainly undergo a radical change as it is integrated into a larger urban system as a very specialized part. For example, local trades and services -- the local downtown -- could become revitalized by the increased "foreign" earnings of the commuting residents. But more likely that same transportation improvement that speeds workers to plants on the edge of the metropolitan area also turns local households more to distant regional shopping centers, down that same improved highway. A serious loss (inter-area transfer) of local trade and tax base follows. (The parallel with the effect of urban expressways on downtown business in the large central city is noteworthy.)

A second problem, derived from the first, is that members of the local labor force that previously walked or rode buses to work in the old downtown or former local factory district will not be rescued by job opportunities forty miles away and will keenly feel the decline of local trade and service work, not to mention the coupe de grace administered to the local bus system by the new outer-directed personality of the town.

Finally, an old and aging dormitory satellite runs the risk of just wearing out. As the existing stock of dwellings age and filter down to lower income households, this dormitory satellite will become a low income enclave, unless new houses are injected into the filtering process. But with the gradual disappearance of the merchant and professional class there may be very few left in town who can afford new houses. A declining residential tax base may therefore be joined to the loss of the non-residential tax base, and a blighted town with a starved public sector is more likely to be avoided than saved by the coming tide of exurbanites flowing out of the nearby metropolitan area. "Rescuing" remote small places with highway improvements, as conscious public policy would have to be well handled to avoid simply creating North American _barios_.

Populating Empty Areas

Finally, for the most part, remote small towns will not be saved, they will be replaced. The exhortation that future generations be not deprived of small town living as a possible choice need not be heeded by preserving old mining towns on scarred hillsides or weathered farm villages on bleak plains. We have in fact long been creating replacement towns in new forms: retirement communities, vacation villages of second homes, artist colonies and professional and scholarly retreats. This swelling movement toward the "empty areas" has, however, caught us with little policy and not much more thought. The routine (mindless?) replication of traditional (discredited?) suburban forms threatens to repeat past mistakes in land planning -- "sprawl" and heavy reliance on the automobile -- and seasonal occupancy does not lend itself to the development of competent government, at least in traditional forms.[4]

Some have argued that the empty areas will come to be populated or re-populated as the principal residence of executives and professionals who because of growing "telemobility" can have their desks anywhere and prefer to have their homes in natural environments. Populated by consultants, thinkers, writers and artists, these new-style "rural" areas could come to exhibit higher average incomes than cities -- "income inversion."[5] Such an income trend would be reinforced by the continuing elimination of marginal farmers and the accelerating professionalization of agriculture. Federal rural development policy would then become very different from that now pursued or contemplated. At the very least, a strong land policy would seem to be critical if we are to extend the options now enjoyed by the few into the future under the pressure of the many. Choosing to live in selected places (e.g., Southern Oregon) could be taken as implied consent to live differently: restricted use of automobiles and trailbikes, reduced waste-making, narrower limits on the alteration of natural land forms and so forth Fragile minority life styles would have to be protected from majority rules.

Growth and Migration

The Rate of Local Growth is the Change in City Size

A national policy on the size distribution of cities that does not also incorporate a position on the rate of growth of local populations, that is, the rate of change in city size, would soon be obsolete. One might, in fact, argue that local well-being is more immediately sensitive to rate of growth than to city size, and that the "proper" pattern of city sizes might better be derived from the preferred pattern of local growth rates. But, while there have been many studies of actual urban growth patterns, the literature has little to offer on the optimum rate of local growth in population, or even the analyst's own preferred rate of growth.

One does not have to defend the existence of an "optimum" rate of local growth to argue that growth can be too slow or too fast to be easily or well assimilated. A strong case can be made for local growth in employment and population at about the rate of natural increase, or roughly the rate of national increase. An average rate of employment growth has the advantage of avoiding the chronic unemployment and underemployment that accompanies slow growth in the demand for labor and/or the debilitation of the population stock that follows from "corrective" out-migration (of the younger and more educated workers). In the other direction, average growth avoids the chronic shortages and congestion that accompanies rapid growth and heavy net in-migration, following from the typical lag of the local public sector in adding to the supply of streets, sewers, schoolrooms and trained personnel.

The case for the "U" shaped curve is almost always an easy one to make -- the pathology of the extremes -- but what we need to know is not only where the peak (trough) occurs but also how flat it is around that point, that is, whether moderate deviation from that peak is important or not. On the location of the peak of the growth-welfare curve, a second approximation to this function would bring in population size as a co-determinant. Some have argued that middle-size places of between 200,000 and one million population have the best of both worlds: scale enough to attain moderate economic stability but still small enough to avoid difficult-to-manage complexity. If so, smaller places would be better off being larger, and the sooner, the better. Smaller places would then attain their maximum welfare by growing a little faster than their natural rate of increase, that is by suffering a little extra congestion or shortages for greater scale and choice sooner. Urban areas of over one million population should then trade a little adverse net out-migration -- loss of youth and talent -- for some additional time in which to adjust their much more complex systems, and especially their ponderous (even if sophisticated) public sectors. The growth-welfare curve would then tend to peak a little to the left of the natural rate of increase for metropolitan areas of over one million population and smaller urban places would peak to the right (i.e., at higher than natural rates).

A third approximation of the "best" local growth rate would distinguish between the pull of rapidly expanding employment opportunities in the local economy and the push of unemployment or underemployment in the hinterland -- or better its "migration-shed." Both the local and broader national interest are served, under demand-pull, by local growth at an average rate and at least the national interest is served by an even higher rate of growth, that is, by heavy net in-migration. (It is surprising how often localities act as if they believe that rapid growth will redound to their benefit in almost every way, and how little they appreciate that growth is not a universal solvent but that it produces about as many problems as it solves -- housing shortages, traffic congestion.) The "too rapid" growth of midwestern cities in the postwar period was probably in the national interest in light of the great need to de-populate the farm belt, and the fact that this very rapid growth was based on the strong demand-pull of a great backlog demand for durable manufactures also acted to ease the strain. And yet these cities still show the strain of that boom. In sharp contrast, the above average growth in population in Albuquerque shows more signs of being an under-employment-push from the hinterland, and it is not clear whether the local or national interest is being served by this above average growth. This is further complicated by the fact that Albuquerque is somewhat too small to be a strong regional developmental pole and, other things equal, should trade some congestion for greater scale, but it needs demand-pull growth instead.

In assessing the welfare characteristics of the rate of local growth, we must always take care to distinguish local from national welfare; a given locality can enjoy gains that beggar others. For example, a very rapid rate of local growth could swamp the local cycle, with a cycle trough barely if at all lower than the preceding prosperity peak. A local recession is then expressed as a reduced rate of in-migration, with local employment holding steady.[6] But since total national unemployment is unchanged, one local economy escapes unemployment by exporting it to others. Similarly, rapid growth draws in the most mobile part of the labor force and gives to that local labor market an unusual power to adapt to rapid contraction at some later time, if need be, with quick and easy out-migration.[7] But again the nation might be better off if the more mobile elements of the population were more diffused throughout the country, minimizing the maximum risk of structural unemployment. Rapid growth also up-dates the local capital stock, making it newer and prettier (to many) and attracting those sensitive to (or guilty about) seeing blight. And again this can be a beggar-thy-neighbor gain; central city mayors have fought New Towns primarily because they provide a too easy escape from the depressing effects of poverty and blight.

The gain to the nation from very rapid local growth lies almost wholly in the realm of traditional industry or consumer-oriented economics. A sudden large increase in demand for a product or service produced in only a very few places forces a rapid expansion of those places so that the consumer (the nation) is not kept waiting longer than absolutely necessary. When would the total national welfare be advanced by keeping the nation waiting a little longer in order to build cities a little slower and a lot better?

Migration, City Size and the Market

Can we trust the market and self-interest to adjust the size distribution of cities through migration? On the whole, social scientists have preferred to work with the process of migration because it is a clear act, observable and measurable in a familiar behaviorial framework, rather than with the structure of city sizes, the culmination of a long and complex chain of historical, physical, technological and institutional forces. But migration is much more complicated than it appears on first impression, for there are many less-appreciated linkages between persons to add to the more-appreciated non-market forces.

Small places do not empty out as promptly or as fully as they would under a pure market model of behavior for at least two reasons. First, each wave of out-migrants draws more than proportionately from the more educated, talented and ambitious elements of the local population, leaving behind an ever weaker labor pool from which to draw the teachers, counsellors and leaders of all kinds that must meet the challenge of re-working a harder and harder core of unemployables and immobiles. Reinforcing this adverse sequence is the conflict of interest between parent and child. Middle-aged parents with poor schooling and few, if any, job skills that will transfer to the newer, larger place are often better off staying on the farm or in the village through the remainder of their working lives, and on into retirement. But their children have no future in agriculture or in the small place and, in fact, face a bleak future anywhere if they remain in the local school system through graduation. Inter-generational ties and a conflict of interest, with decisions made by unsophisticated parents, is clearly non-optimal from a social point of view.

At the opposite end of the city size spectrum, very large urban areas probably do not slow in growth as much as they should under the pure market model, because migration is characterized by a decided asymmetry. Large cities appeal more to the affluent because they offer much greater consumer choice, especially to the more educated element seeking the more esoteric offerings, and to the young adult with a thirst for adventure and an unlimited faith in his (her) potential to win big in the biggest arenas. But these are also the more mobile persons, so that those who prefer big cities tend to move to them. With time and aging, the preferred life styles tend to shift more in line with the environment of smaller places, but the elderly tend not to move easily due to sunken investments in homes, friends and local institutions and due also to the shorter remaining life over which the cost of moving must be recaptured.

Note that one can hardly keep from "choosing" to live in a larger place; one only needs to not move. A person of retirement age could have lived, as an adult in command of his own location, almost five decades in the same city and have witnessed, even at a moderate rate of growth of 20 per cent per decade, a population increase of 150 per cent -- from, say, 400,00 to one million. There is an "age-bias" in migration toward bigness because those who prefer large cities do tend to act on those preferences and those who prefer smaller places tend not to act. Note also that this age-bias tends to reinforce the "skill-bias" in migration, referred to above, through which professional and technical workers lock the semi-skilled production workers into their locational preferences for larger urban places.

27 May 1971

On Containing the Size of the Largest Cities

Slowing In-Migration versus
Promoting Out-Migration

Perhaps the reason that we do not seem to be able to come up with an operational strategy for containing the growth of very large cities is that there isn't any. And we should have been addressing ourselves instead to improving their internal organization. Most of the emphasis in the movement to place an upper limit on the size of cities is on discouraging newcomers, but the simple arithmetic of the situation establishes that the problem is more that of inducing native residents to leave town. Natural increase accounts for about six-sevenths of the growth of metropolitan areas of over two million population.[8] It seems most unlikely that we can much slow the natural gravitation of young adults to the biggest places, and unnatural to try to turn them away. There are, however, some unexplored possibilities for increasing gross flows <u>out</u> of big cities.

A heavy rate of out-migration that reduces local population increase below the rate of natural increase has always been a cause for concern because it tends to be a "youth drain." But this problem almost always arose in a small town context; big cities typically experience net inflows of young adults. Besides, it is the middle-aged and the elderly who are ordinarily the most willing to leave big cities (even if not the most able), with the former oriented more toward middle-size places with good jobs and schools, and the latter looking more to smaller places in which it is easy to get around. Retirement communities are, in fact, much over-represented among the nation's fastest growing urban places. How much has the growth of these places already slowed the growth of big cities, and how much more of a redistribution of population could be achieved with a more intensive promotion of retirement communities?

Transportation Technology and Natural Ecology

But even if we should begin to approach "zero population growth" as a nation and zero net migration into metropolitan areas, we must still reckon with the horizontal spread of cities. Every transportation advance that increases speed of movement, and thereby the distance that can be traversed in the generally accepted one hour travel time limit, extends the effective radius of the metropolitan area and encloses more land area. The smaller metropolitan areas may often be surrounded by sparsely populated hinterlands and can perhaps be clearly delineated in boundary and precisely defined in internal growth rate. These smaller metropolitan areas tend, in fact, to grow by drawing population out of their hinterland. But the largest metropolitan areas have so great a power of attraction and generate such strong "spread effects" that they tend to be surrounded by densely populated hinterlands. The effective New York and Chicago urban regions will literally over-run scores of previously independent towns and small cities over the next decade. From independent city to satellite to contiguous suburb in a decade or two.

A forty per cent increase in the average commuting speed (radius) --a not unreasonable expectation for the year 2000-- would double the land area of the local economy. If, moreover, progress in telecommunications reduces the need to interact personally with others in central places to one-half of the present level, and if the work week has been cut to three days by then, most of us would need to "commute" to a large center only once or twice a week. A resident of a suburb or satellite, sixty or eighty miles from downtown Chicago would probably visit there only twice a week, for all purposes, business and recreational. Whether these "commuters" should be included in the Chicago area growth and size statistics is an academic point; the critical issues lie more in the land planning of these far-flung, but economically integrated, urban regions. The growth rate and population size of our largest metropolitan areas, taken in aggregate, not only can not be defined, they seem not even to be meaningful concepts. Witness those tortured statistical artifacts that the census has had to construct: The New York-Northeastern New Jersey Consolidated Metropolitan Area and The Chicago-Northern Indiana Consolidated Metropolitan Area. If transportation progress is inexorable, larger city size seems inevitable.

Intra-urban land-use planning will, of course, come to revolve more and more around serious considerations in natural ecology. Criteria of density and spacing must include but go well beyond amenities of open space, aesthetics of landscape architecture and convenience of movement, into the new world of waste disposal and pollution and the re-cycling of matter. Physical planners have, of course, long been concerned with the loss of open space in and around cities. But too often they over-played the "calamitous loss of irreplaceable prime agricultural land" which in some not too clear way was held to threaten our very existence. Their anguish in a time of farm surpluses did not transfer well to their critics.[9] Tax abatement for near-in farms never did elicit much support, nor would it have prevented many rural to urban land conversions. Care would have to be taken that property tax reductions did not simply subsidize and encourage land speculation. The case for urban open space has always seemed to rest primarily on recreational use and secondarily on its use to delineate boundaries more sharply to strengthen community identity and build civic responsibility.

An ironic turn of fate may bring forth a new variation on the old theme of agricultural land as urban open space. Experimental design is underway in at least one urban area (Muskegon, Michigan) for dumping liquid household effluent on porous soils (sand dunes) over a period long enough to enrich these soils to good agricultural quality. Waste disposal becomes the re-cycling of matter.

Natural ecological imperatives could come to dictate population densities and urban land-use patterns. Given then the speed of transportation and the maximum acceptable commuting time (one hour?), the urban area may come to be defined "from the ground-up." The current direction of causation would then be reversed: city size would not determine population density and land-use patterns but rather land-use patterns and population density, as bounded by current transportation technology, would determine city size. At least in the formal accounting, the population continuum across space could be broken every hour of travel time and summed into (trivial?) population aggregates if one just had to know summary (and nominal) growth rates and city sizes.

27 May 1971

On Industrial Structure

From Nation-Wide Industry Location
to Local Industry-Mix

We have a rather considerable literature on the efficient location of industries, in partial equilibrium, but only scattered sentences on the optimum industry-mix for an urban region, especially in a longer time perspective. The distinction between the industry and regional vantage points can be visualized by imaging a series of industry locational maps, one for each of the many industries, overlaid one upon another, through which a vertical shaft is sunk at some given point; the "core" so removed would then be labelled Cincinnati or Phoenix. Even if each industry were arranged in space in an efficient way, would the many layers yield "cores" of industries that were optimal, or even viable, at each of the many localities that sum to make up the national economy? Would we not run the risk that some places would have too many jobs for men and too few for women (Wheeling-Steubenville or even places as large as Pittsburgh) and the reverse (Carolina textile towns)? Would we find high-wage but cyclically unstable durable goods centers in need of more non-durables or services for stability, just as the non-durables towns lack for the heavier, higher-wage durables? And places with sharp winter peaks, unrelated to local climate, posed against the reverse?

There is good reason to doubt that the market will adjust any and all such distortions with compelling price signals and irresistable corrective processes. True, textile mills and garment shops did move into coal and steel areas to tap the large pools of redundant (cheap) female labor, but there are some very clear weaknesses in the connecting chains. The blending of the capital intensive ("heavy") industries, invariably unionized, with the lighter work of the more mobile ("footloose") industries would not seem to happen naturally or automatically through the market forces. Typically, oligopoly product price power in combination with aggressive unionism raises heavy-industry wage rates which then "roll-out" into the other occupations in these high-wage centers. There is scant reason to bring the lighter, lower-paying, non-unionized operations into that area, except perhaps to serve that local market if it is unusually large. Similarly, there is little incentive to locate the high-wage work in the low-wage area because the union wage goes right along with the move and the economies of agglomeration and urbanization of the larger center often do not follow. To the extent that these "heavier" and "lighter" industries generate complementary labor demands -- male and female, skilled and unskilled, heavy and light physical requirements, offsetting cyclical or seasonal patterns -- private cost-minimizing location that separates them has important social costs in income inequality through unemployment and under-employment.

Industrial Filtering in the System of Cities

The hypothesis has been advanced elsewhere that industries filter down through the national system of cities. Invention, or at least innovation, takes place more than proportionately in the larger urban areas of the more industrially mature regions, and as industries age and their technology matures, skill requirements fall and the industries become free to relocate in lower wage (lesser skill) areas. The higher a city stands on the industrial skill hierarchy, the younger its industries and the more likely it is to fashion an average rate of growth out of a fast-growing industry-mix and declining shares of that work -- the innovation of new work and the spinning-off of work that has become routine. The lower an urban area in the skill and wage hierarchy, the older an industry tends to be when it arrives in town and the slower its national growth rate. Intermediate level places tend to fashion a slightly above average rate of growth out of growing shares of slow-growing industries, but below this size the positive change in share weakens and erodes to zero, leading to slower than average growth (net out-migration) below about 25,000 population and absolute employment decline below 2,500 population.

This hypothesis has not been subjected to anything more than very preliminary testing (which did support it) and probably can not be rigorously tested without a more disaggregated data than the U. S. Census S.I.C. two-digit ("industry group") employment figures that currently serve as the basic data of "shift-share" analysis. But the credibility that even casual reflection and observation lend to this hypothesis does seem to suggest that it would not be premature to give some thought to the implications that a national industry filtering process carries for national policy on the distribution of population. Do we now or can we learn to assimilate well, at the national level, such a spatially discrete hierarchy of skill (education), wage level (income) and employment growth (vocational opportunity)? Can this dynamic variation on the more static theme of central place theory be accepted with mild interest as a theoretical or empirical curiosity, as seems to be true of so much of regional economics, or are there important normative and policy ramifications here?

If industries do in fact filter down through the national system of cities, then human resources would tend to filter up in a complementary way. The textile and apparel towns of the Piedmont area have been able to maintain full employment of a rapidly expanding labor force -- have been able to absorb the exodus from agriculture in that region -- by capturing an ever larger share of these slow-growing industries. But tight local labor markets have not produced an average per capita income, as the more ambitious and talented young adults of that region have migrated out (filtered-up) to various larger places. Worse still, this trading of high talent for low skill work has compromised the long run development potential of that region.

This is, moreover, a double-edged sword in that the larger, more industrially sophisticated urban areas of the North struggle with heavy unemployment that leads to near unemployability and, in large measure, for lack of low-wage unskilled work. New York City needs low skill work as desperately as a textile town needs skilled work. In general, if natural increase tends to produce a population with a random distribution of talent and ambition, and if industrial filtering tends to sector high and low skill work into distant local labor markets, then massive migration is dictated. But if the more talented, motivated and educated are the more mobile, the net flow is biased toward the larger, higher skill places.

Large Cities as Instruments of National Economic Policy

The urban hierarchy and the industrial filtering process, as presently constituted, could perhaps be assimilated in a way which would serve the national interest if we were to act to increase the mobility of labor, especially the lesser skilled workers, and to arrange inter-regional fiscal transfer of a magnitude that would ensure that local public services were made reasonably uniform from place to place, to assure equality of opportunity. Or we could instead see in the current trend toward large metropolitan areas the resolution of the problem of balancing local labor markets, and reinforce that trend. If nearly the whole national population were contained within a handful of very large metropolitan areas, each of which would perhaps be more a conjuncture of overlapping local labor markets than a single, indivisible commuting space, then virtually the full range of occupations would be within easy reach of nearly everyone. Every area would not, of course, produce every product; inter-regional product specialization and trade would continue. And gross flows of migration would go on, as between universities for example. But a nation of multi-million population metropolitan areas would produce balanced local labor markets as a by-product.

27 May 1971

Such a distribution of national population would also greatly simplify national economic policy in many ways. Expansionary monetary and fiscal policy is complicated when significant inflation begins to appear in some places (with leading industrial sectors) before full employment is attained in other places (with lagging sectors). Similarly, repressive monetary and fiscal policy can create serious unemployment in, say, durable goods centers while strong inflationary pressures persist in service centers. Variation in local business cycles can not be fully explained by differing industry-mixes but the industrial diversification that accompanies larger size would certainly make an important contribution to inter-regional symmetry in the response to national economic fluctuations and their treatment. (It is only fair to point out that the inner city ghettoes of our largest metropolitan areas would have to be integrated much better into the regional economy of which they are only nominally a part now before we could make a fully convincing case for the employment stabilizing virtues of the very large local economy.)

Again, the industrial diversification characteristic of large metropolitan areas tends to generate similar income patterns, and thereby reduces the need for heavy inter-regional fiscal transfers. Federal policy-makers are then freer to concentrate on programs or incentives that induce a more efficient land-use pattern or that create a more effective organization of local government, if they were less bound by equity considerations or income goals. It is much easier to deny an average, rather than a low, income area federal funds if they balk at raising pollution standards, consolidating services or socially integrating populations.

Related to the previous point but worthy of special mention is the tendency for industrial diversification through the concommitant convergence in educational, occupational and income characteristics, to significantly reduce the inter-regional spillovers that occur through migration. If large metropolitan areas come to trade populations with very similar socio-economic profiles, a major source of unintended inter-regional redistribution of income and wealth would be averted, such as occurs when small towns give up the top of their high school graduating class and take on responsibility for displaced rural migrants with only a few years of poor schooling. All in all, a nation of big cities greatly simplifies Federal policy, or at least it would if these large *economic cities* were internally competent and effective as local governments. If we do continue moving toward becoming a nation of very large metropolitan areas (economic cities), then the cutting edge of public policy will be less to find an optimum system of cities and more to create an optimum system of governments.

On Locational Patterns

The Seeming Drift to the Coasts

There are two dimensions of the spatial pattern of urbanization in the United States, at the gross grain of the nation as a whole, that stand out over all others: the emptying out of the center in the drift to the coasts, and the physical formation of great linear strips of cities -- "megalopolis." The drift to the coasts has been rationalized as a seeking, in part, of easy access to water recreation, a reflection of both a rising per capita income (through an income elastic demand) and a change in taste patterns. More recent in nature, the surprising increase in our imports of foreign manufactures, together with the substantial technological progress in water transportation, probably acts to strengthen the position of seaboard cities. Port sites are, of course, especially favored in the processing of intermediate products where further manufacturing is required.

Some caution should be exercised in a too easy acceptance of the drift to the coast as an independent factor in the American population distribution. About half of the coastline is not experiencing unusual growth; the South Atlantic Coast has, in fact, lost share of population for three decades now and has exhibited absolute de-population over much of its length. Again, one still does not encounter significant urbanization north of San Francisco short of the Puget Sound region, eight hundred miles away. What appears to be a drift to the coasts is in great measure a simple reflection of the growth of large cities located on the coasts -- the momentum of prior urbanization. What is perhaps more defensible is the proposition that a large population (the scale effect) and location on the seaboard or the Great Lakes are sufficient conditions for rapid population growth to large size and substantial growth beyond even from a large base. But apparently not necessary conditions: see Phoenix and Denver.

Megalopolis Revisited

The great urbanized strips that have spread across the country -- "megalopolis" -- are also probably in part real and in part apparent. Gottman and others have commented at length on the powerful externalities that characterize the Boston to Alexandria strip.[10] But perhaps equally important in its formation is the simple historical circumstance that the nation was colonized along the East Coast and settlements that were more than a day's journey apart at that time -- "separated" -- are less than an hour apart with today's transportation -- "joined." No one has compared the current population pattern in these urbanized strips with that which would have resulted if each city had grown for two centuries at the normal (national) rate for that size city. How different is the observed settlement pattern from the sheer momentum of urban growth, taken in isolation?

In any event, strips of cities have the special property of combining a relatively high degree of access to a variety of jobs and goods with relatively close open space. This land form produces overlapping local labor markets; more workers live within commuting range of two of more employment centers. In so far, moreover, as narrowly channelled movement speeds movement through more capital intensive transportation systems, this further increases occupational and consumer choice in space. Because mobility increases with income, in many ways, one gets the suggestion of a partial resolution to the problem posed in the opening pages. Strips of physically distinct middle-size cities, drawn together in time by very rapid (if sometimes expensive) transportation, might happily offer an escape from physical bigness to some and access to economic bigness to others. And with proper land-use control, open space for recreation can be preserved nearby, perpendicular to the strip. By comparison, small, scattered towns gain closer open space but suffer thinner local labor markets, and large circular cities mass jobs and shops but lose access to the countryside. Strips of cities may not be all that bad.

Low Density Urbanization

A third spatial dimension of American urbanization and one that seems destined to become a major policy focus is the steady conversion of high-density agricultural land into low-density urbanized regions, ranging from the more heavily populated Piedmont Crescent of the Carolinas to the less populated delta of Eastern Arkansas. The Piedmont population, in aggregate, rivals that of the Atlanta city-region, but it is still an open question as to whether its loose-knit form can generate as much developmental power as is characteristic of that more classic urban form. The Piedmont has made a relatively successful transition from primary industry into semi-skilled manufacturing, but the next bigger step to science-based manufacturing or to exportable professional services is in doubt. And this second, more demanding step, must probably be made even more quickly under the pressure of the growing shortage of semi-skilled and skilled manufacturing work, as successively more sophisticated operations are being spun-off to industrializing nations all over the world. If this sprawling system of small to medium size cities -- the largest is only a little over one-half million population -- can not learn to conduct research, marshall information, make good decisions promptly and sell itself in a "post-industrial age," a second major population re-adjustment will become necessary, dwarfing the current agricultural de-population.

There are half a million people in "Eastern Arkansas," described as a sixty mile radius circle drawn around an imaginary center. If that population mass could function as an urban region, it would be a middle-size metropolitan area, and belong to a class that has exhibited the strongest rate of population growth, better than national average income characteristics and a relative domestic tranquility. The growth of automobile expressways in this and similar low density urbanized areas has brought about a loose integration of what had been a multitude of local labor markets, at least for the more skilled workers who can afford automobiles. The heavy cross-commuting of such an area places it in some uncertain middle ground between its former life as many isolated local economies and its potential future as a single integrated economy.

But this population is not at all comparable in socio-economic status to that which normally has been assembled in a metropolitan area of one-half million population. Eastern Arkansas holds today a residual population: a few who have made it in agriculture (large scale, capital intensive farming) and many more who have not, and a trade and service sector that has grown up relatively sheltered from competition.

There are other notable differences between the classic city and an agricultural area that has weakly and tentatively urbanized. Old houses are scattered far and wide and many (most?) are not served by any form of public transportation, except the school bus. The normal filtering down of housing that demands no more of a low income family than that it move another mile farther out from the old city center, along the same old bus route, here requires that the family change towns. Psychologically, this is probably more a migration than a "move." A corollary to this is that "towns' come to be abandoned, like old neighborhoods in the large city, as part of the normal process of residential filtering -- towns like houses may simply wear out.

Many of these old hamlets and villages should not be recycled with new housing for they are old rural service centers that have long lost their economic function, and have not provided local employment for years. Their continued existence can often be justified only to the extent that they relieve the housing shortage of the region, but they perform this service only at the cost of isolating the families forced to inhabit them. This is an "awkward age" that must be bridged on the way to some new land-use pattern that relates to some new transportation system, such that a full range of housing -- houses of all ages and quality -- is arrayed along the principal channels of movement. An evolution from dispersed to linear residential patterns does come easily to mind but other options may also come to mind after harder thought. The principal point to be made is that we must learn how to interconnect half-a-million people, spread out over a dozen counties, to achieve viable economic scale or we must assimilate them as migrants.

27 May 1971

Reprise: Migration in a System of Cities

A number of ideas that seem to dominate the discussion of population distribution policy can be brought together usefully in a simple schematic figure and summarized as follows:

a) The de-population of rural areas should, at least on theoretical grounds, lead to higher returns per worker and higher levels of money income and well-being, as shown by point A. Logically, rural areas should be left with not only fewer farmers but also the best farmers.

b) If rural out-migrants were to locate in middle-sized urban areas (D), they would tend to increase the money income (productivity) and probably further increase the real income of the inhabitants, as the growing local market permits greater range of choice in goods, services and occupations. If the place of out-migration is in the stage of diminishing returns and the place of in-migration is in the stage of increasing returns, and wages (income levels) are by implication higher in the latter place, everyone benefits: those who move, those left behind and those being joined. Migration here (from A to D) is clearly in the public interest.

c) The most discussed case of the day is the migration from small towns (B) to very large cities (E). Such moves usually benefit the migrant who rises from welfare level B to E but could leave everyone else worse off. Those left behind in the small town, growing smaller, face higher costs of utilities and higher taxes for those hard-to-contract (indivisible, heavy-fixed-cost) operations, and a reduced range of choice of goods, services and

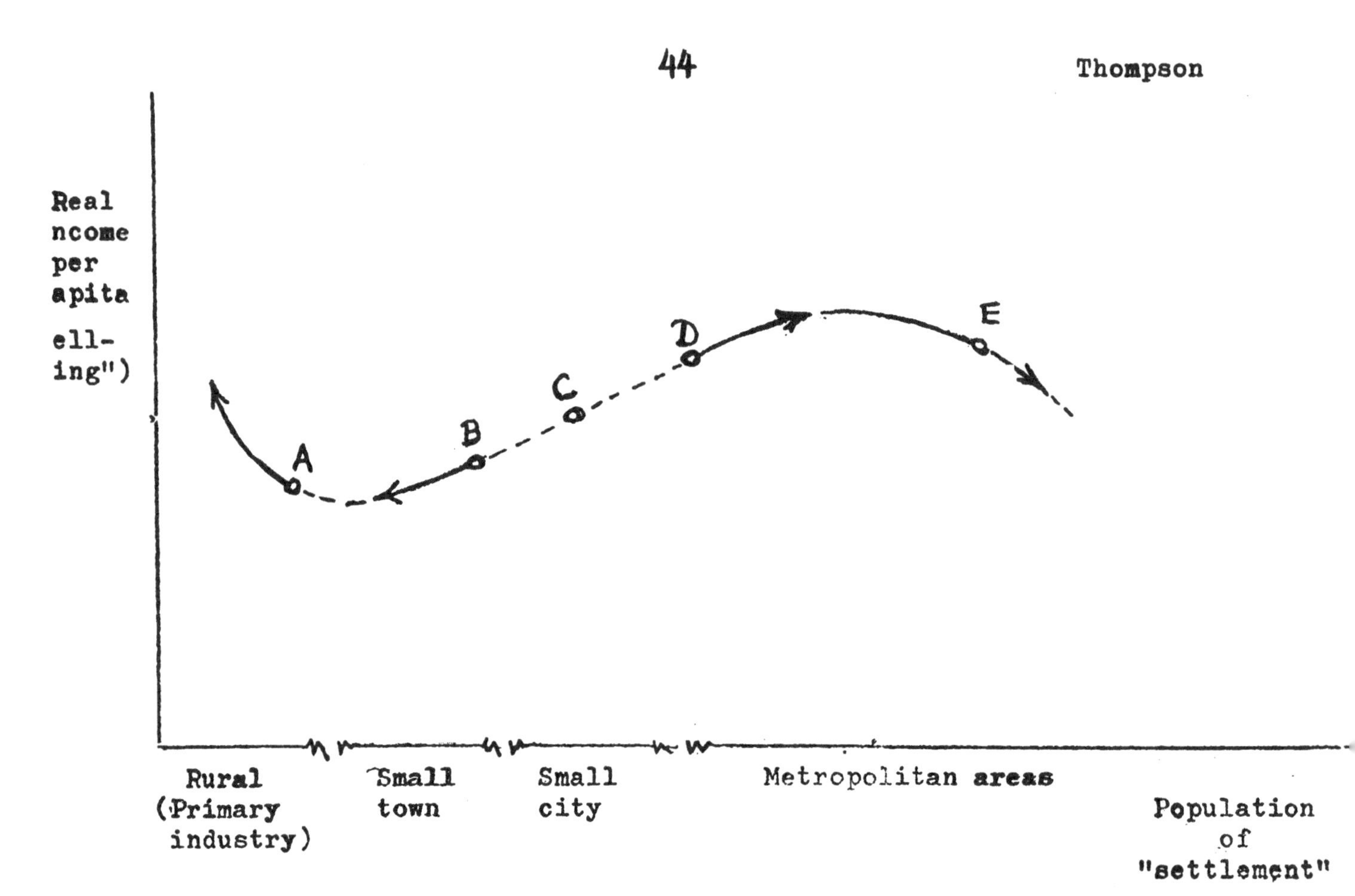

Schematic Representation of the Relationship Between the Welfare of Migrants, Those They Leave Behind and Those They Join, with Special Reference to the Related Size of the Various Populations.

27 May 1971

occupations. The big cities into which the migrants move are also burdened with increased congestion, greater trip distances, more political fragmentation and inner city housing shortages. If this is the case, then the price signals are misleading; we have made it too cheap to leave small towns and/or too cheap to enter big cities. Migration is profitable for the individual but not in the public interest.[11]

d) The population drift from small towns to big cities has lead some observors to argue for government intervention designed to divert the migrants more toward middle-sized places -- small metropolitan areas of 200,000 to one-half million population.[9] These healthy, fast-growing places have demonstrated their attractiveness, vitality and viability, and may also still be in the stage of increasing returns. A larger population will improve their university, museum and theatre offerings and public transportation system. Movements from small towns (B) to middle-sized cities (D) improve the positions of the migrants and those they joined, but do leave those left behind worse-off. And the growing gap between incomes in declining places of 10,000 or less and rapidly growing places of 100,000 or more will induce further migration. A cumulative disequilibrium process is set in motion which seemingly can only end with the abandonment of remote small towns (B).

e) Perhaps the most popular urban development policy of the day is the "growth center" or "growth pole" strategy. While there are important variations between countries, the domestic (Appalachian) version most often argued is designed to identify "small cities" that have growth "potential" but may or may not develop depending on chance, or may develop unnecessarily slowly and painfully. If, then, migration from rural areas (A) and small towns (B) were coaxed into these marginal places (C), risks and/or delays in their development would be reduced. Such a policy is inferior to the one immediately above in that the former would divert migrants to larger places subject to less risk of failure -- investing in smaller potential growth centers could result in throwing good money after bad. In addition, the immediate returns to the migrants will be lower if the wage rate in the growth center of 20,000 population (C) is lower than in the small metropolitan area (D), as is assumed in the curve depicted.

This latter policy is superior to the one above in that, if successful, significant amounts of sunk capital will be saved and social dislocation avoided, as is generally appreciated. What is not explicitly recognized is that the level of well-being (real income) experienced in small metropolitan areas (D) is also a function of their rate of growth in population. And if these middle-sized places are already growing faster than their natural rate of increase in population and must thereby accommodate heavy flows of in-migrants, accelerating that rate of growth may lower real income -- housing shortages, schools on half-day sessions and traffic congestion -- due to lagging public investment. Besides, why hurry these favored places through their golden age?

Footnotes

1. For a recent survey of findings on economies of scale in urban public services, covering both U.S. and U.K. experience, see Harry W. Richardson, Regional Economics (New York: Praeger Publishers, 1969) p. 195f.

2. Committee for Economic Development, Reshaping Government in Metropolitan Areas (New York: CED, 1969)

3. Niles M. Hansen, Rural Poverty and the Urban Crisis (Bloomington: Indiana University Press, 1970)

4. Dennis Durden, "Use of Empty Areas," in F. Fraser Darling and John P. Milton (editors), Future Environments of North America (Garden City, N.Y.: The Natural History Press, 1966) pp. 479-93.

5. Brian J. L. Berry, "The Geography of the United States in the Year 2000," Ekistics, Vol. 29, No. 174 (May 1970) pp. 339-51.

6. Wilbur R. Thompson, A Preface to Urban Economics (Baltimore: The Johns Hopkins Press, 1965) pp. 176-77.

7. Peter A. Morrison, "Urban Growth, New Cities and 'the Population Problem'" prepared for the American Association for the Advancement of Science, December 29, 1970, Chicago, Illinois, p. 24.

8. Conrad Taeuber and Herman Miller in a statement before the Ad Hoc Sub-Committee on Urban Growth of the Committee on Banking and Currency of the U.S. House of Representatives on June 3, 1969 in Population Trends, Part 1 (Washington: USGPO, 1969) p. 4.

9. Mason Gaffney, "Containment Policies for Urban Sprawl," Approaches to the Study of Urbanization (Lawrence: University of Kansas Press, 1964)

10. Jean Gottman, Megalopolis (Twentieth Century Fund, 1961)

11. Calvin L. Beale in Population Trends, Part 1, op. cit., pp. 473-508, and James L. Sundquist, "Where Shall They Live?" The Public Interest (February 1970) pp. 88-100.

27 May 1971

May 6, 1971

Optimality in City Size, Systems of Cities and Urban Policy: A Sceptic's View

H.W. Richardson*

NOTE: This paper has been prepared for the Resources for the Future-University of Glasgow Conference on Economic Research Relevant to National Urban Development Strategies, Glasgow, Scotland, August 30-September 3, 1971. It is subject to revision and should not be cited or quoted without express permission of the author.

H.W. Richardson, Director, Centre For Research In The Social Sciences, University of Kent at Canterbury, Canterbury, Kent.

THE NATIONAL SYSTEM OF CITIES AS AN OBJECT OF PUBLIC POLICY

OPTIMALITY IN CITY SIZE, SYSTEMS OF CITIES AND URBAN POLICY: A SCEPTIC'S VIEW*

I INTRODUCTION

In developed economies the share of agriculture and other primary products in GNP is generally so small (and declining) that it is excusable to think of economic growth in terms of urban growth. If so, we may treat the national system of cities as the spatial form of organisation adopted by the industrial economy to achieve its growth goals. This adds another dimension to discussions of growth in the aggregate economy, and the spatial-urban dimension is fully as worthy of study as the more familiar macroeconomic and sectoral approaches. As in the latter cases, the question of optimality figures prominently in the spatial-urban context. Is there an optimal city size and/or an optimal distribution of city sizes? How true is the widespread belief that the large city makes a much greater contribution to national growth than to the welfare of its citizens? What are the implications of findings on these questions for formulating a national urban policy?

Since the sparse literature on this subject is littered with inconsistencies and confusion, I make no apology for the fact that this is a conceptual and discussion paper rather than the results of hard research. Clarification of some of the issues arising out of the existing state of knowledge is an important prerequisite for worthwhile future research. This paper falls into three parts: Section II, the core of the paper, deals with the concept of optimal city size; sections III-V reviews some of the literature on the national urban hierarchy; and sections VI-VIII looks at the goals and instruments of urban policy, and explores the policy implications of the earlier arguments.

II OPTIMAL CITY SIZE

An integral aspect of the problem of an optimal distribution of cities and whether or not it is feasible to influence this by government action is the question of whether there is an optimal city size, or even an efficient range of sizes. This question is now arousing increasing interest, but deriving firm conclusions is made more difficult by the sparsity and poor quality of the available statistical evidence. However, it will be shown that even from a theoretical viewpoint

* I wish to thank my colleagues Murray Stewart and Joan Vipond for comments on an earlier draft of this paper.

the search for the optimal city size is unsound. There may be more sense in trying to identify efficient ranges of city sizes between a minimum threshold and a scale at which further increases in size are not accompanied by additional agglomeration economies. Even so, we may expect the efficient range to vary, possibly dramatically according to the functions and the structure of the cities in question.

Much of the early work on the optimal city size problem concentrated on examining how the costs of urban government services varied with city population size, and many observers have argued the case for the traditional U-shaped curve and derived the optimal city size from the bottom point of this curve. This is both theoretically unsatisfactory and incomplete. However, even if we accept this approach at its face value it yields ambiguous results. Estimates of the most efficient size range have varied widely: 50-100,000 (Hirsch, 1959); 100-150,000 (Lomax, 1943); 500-1,000,000 (Duncan, 1956); 100-200,000 (Clark, 1945); 30-250,000 (SVIMEZ, 1967 - quoted by Cameron, 1970); 100,000-250,000 (R.C. on Local Government in Greater London, 1960); 250,000-1 million (Redcliffe-Maud Commission, 1969). It is not surprising, however, that estimates should vary from country to country and over time as the components of the urban government service mix and relative cost conditions alter.

Apart from this, the procedure has several drawbacks. The urban government service mix varies even among cities of the same size. The theoretical solution of a 'representative bundle of services of given quality in a standardised city' is unacceptable not only because weighting changes with scale but even more awkward the number of services supplied will tend to increase with city size.[1] Furthermore, most empirical work has looked at how expenditures per head change with increases in population, but this throws light on scale economies only if demand is inelastic; what is really needed are measure of the influence of scale on unit costs o homogeneous output (Gupta and Hutton, 1968). Other difficulties are that the division between public and private services is is arbitrary and can vary according to the institutional traditions of the economy and that some costs in the estimates may be transfer payments rather than real resource costs.

A more fundamental objection is that urban government service costs are only one part, perhaps a relatively minor part, of the costs and revenues that affect optimising decisions in the choice between cities. Very little has been discovered about the relationship of private costs (i.e. producers' costs and consumer costs) to city size, and the evidence is inconclusive (Alonso, 1970b). There is somewhat stronger evidence that productivity (as measured by output per capita) increases with city size, primarily as a result of agglomeration economies. Presumably there is also some point on the curve when it stops rising and turns down again as diseconomies predominate over economies, though this point may not yet have been reached in most cities. If we combine public and private costs into a single cost curve then we can derive both a cost and product curve similar to the cost and revenue curves in the theory of the firm, with the exception that population rather than quantity of output is measured on the horizontal axis. Since we would never expect the optimal position for each and every firm to occur at the same level of output, why should we expect the optimal point in each city to be located at the same population size?

One possibility (not firmly refuted by the evidence) is that the marginal and average product curves rise continuously and that the average cost (= marginal cost) curve is approximately horizontal. In this case there is no optimum. A more critical case is where the product curves rise continuously but where the cost curves are U-shaped. The intersection of the AP curve with the downward falling AC curve represents the minimum threshold size, and the bottom point of the latter is the minimum cost population size. Neither is an optimum. There are, however, several different optima in this model, each reflecting a particular perspective. For the resident the optimal size is that which maximises the gap between AP and AC. For the planner there are two social optima. The first is where MP=MC; here the city maximises its contribution to total output provided that there is no more productive locality elsewhere for the increase in population. However, this is a true optimum only if either there is an unlimited population or all other cities are at the equilibrium (MP=MC) point. The second social optimum is more usual; this occurs at a smaller city size where MP exceeds MC by an amount equal to the opportunity cost of siting the population in alternative cities.

A very different result is obtained when we look at the optimum from the viewpoint of the individual firm. Firms are not tied to a given city but choose from the available range of city sizes the scale of urban area that maximises its own profits. As von Boventer (1970) has shown, the optimal city for a firm is that city size which maximises the difference between agglomeration economies for that firm and the costs of urban services incurred in that city. Since the importance of agglomeration economies varies from one firm to another, so will the optimal city size. Thus, the optimal city size is indeterminate without detailed information about the firm's characteristics. To obtain more objective overall city size optima it is necessary to find out how many firms have their optima at each particular city size, to aggregate their labour forces and translate them via use of population/labour force ratios into an overall city population. However, this suggests the possibility of incompatibility between the city sizes necessary to maximise profits for individual firms and the city sizes implied by the location of optimising firms. The analysis is incomplete because the firm's optimal location assumes that city size is given yet it is the summation of all firms finding their optimum at a particular centre that determines, and might indeed alter, the population of that centre. This is the result of adopting a step-by-step procedure in a problem that requires simultaneous determination.

Of these several optima, the second planning optimum (MP=MC + opportunity costs) is probably the most important for urban policy. But it is hardly practicable. In a free economy where migration is left largely to individual choice it is not possible to redistribute population from one centre to another in the fine marginal increments that might be needed to equate marginal product and costs in each centre.

The existence of divergent city size optima for different interested groups blurs any meaning attached to the concept. There is no reason why the optimum for households and for business firms should coincide, and though one solution might be to regard the planning optimum of the urban policymaker as the super-optimum this can also yield ambiguous results. The reason is that the social optimum size depends upon the goals and objectives of the planners, and these vary depending on the breadth of perspectiveness of the planners (specifically, whether they are national-, regional- or local and upon their priorities as revealed in the ranking of objectives.

Furthermore, many of the costs and benefits of city size are not expressed in monetary valuations as embodied in the cost and product curves. Social costs associated with increasing city size, such as pollution, noise and congestion, are not registered in market prices. Because of these unwanted and, as yet, non-quantifiable externalities a social optimum size may be reached at a point where the marginal product curve is still rising faster than the marginal cost curve. Yet because incoming firms and migrants have to pay only the private costs and not the social costs of locating there, market forces may stimulate continued expansion of the city well beyond this optimum. The situation may be exacerbated by the existence of crude if inevitable pricing policies that result in average cost rather than marginal cost pricing for most urban services.

A second and not unrelated point is that there is no inherent reason why the optimum should be an economic optimum. Duncan (1956) listed all kinds of factors that might be relevant to determination of an optimum: accessibility, health, crime and safety, educational facilities, leisure and recreational activities, social institutions, community and family ties, psychological and mental health hazards, and the desire for political participation. More important than any of these are probably the locational preferences of households in regard to size of place (Hansen, 1970) though these preferences undoubtedly subsume some of the criteria listed above. Hansen has referred to survey results that indicate residential site preferences in favour of mediumsized cities or smaller towns rather than the large metropolises (Hansen, 1970; Gallup, 1968; Girard and Bastide, 1960). How important these non-economic determinants are depends, yet again, on the policymakers' objectives and on the characteristics of the social preference function that these presumably represent. However, once we accept the relevance of non-economic and non-quantifiable criteria we are faced with a weighting problem that is virtually insoluble.

There is a close degree of kinship between the concepts of optimum city size and optimum national population. No serious economist would use the optimum population concept nowadays, and it is paradoxical that urban economists, supposedly extending the frontiers of a new field, should find themselves chasing, if only in an anologous form, such an ancient will-o-the-wisp. The gross defects of the theory of the optimum population are well known: the dependence on marginal productivity theory and the law of diminishing returns, the assumption of a closed economy, the use of a static model in a subject that cries out for growth dynamics, and the use of population as an index of economic well-being. The first is sometimes excusable for analytical and pedagogic convenience. Second, the 'openness' of city economies accentuates the theory's irrelevance for urban economics. Third, a static framework is totally inappropriate since the optimum population concept has been discussed in the context of economic development. Similarly, the notion of optimum city size is not immediately consistent with the framework of a growing system of cities. It is true that the two things can be made compatible, but the results do not make very good sense. For instance, we could treat urban policy as a kind of bottling plant where each city represents a bottle of a standard size and where the policymaker's job is to fill each bottle to the brim in turn then move on and fill the next. This might be a practicable urban strategy but it is most unlikely to lead to an efficient system of cities. A modification of this approach is for urban policymakers to hold up population expansion when the so-called optimum is reached and then to shift attention to increasing *per capita* income. In an expanding economy this will be difficult without interurban migration controls since cities experiencing rapid increases in *per capita* income will also be attractive to potential migrants. Moreover, in a dynamic setting a large city may be functioning efficiently even if the gap (AP-AC) is not maximised or even if MC>MP. This is because in the broader framework of a system of cities the efficiency of a large centre may be related to how well it performs its nationwide functions as a distributor of innovational activity, new ideas and managerial expertise down the city hierarchy.

The fourth objection to the optimum population concept is possibly most serious of all. Optimum city size is invariably discussed in terms of city population. It is unclear whether this is because the level of population is considered to be the most suitable index for determining the optimum (on the grounds that agglomeration economies and diseconomies can be functionally related in a more or less precise way to popultion size), or whether population is used as a surrogate for other size variables for which we lack data. Reliance on population data may be forced upon us by the poverty of urban social accounts, but much of the literature gives the impression that the choice is deliberate. In fact, although useful for some purposes, e.g. determining the minimum threshold 'market' for health and educational facilities, some types of retail trade and public transport systems, population is a poor size proxy in many other respects. Particularly in medium or small cities, the same population could imply very different total urban incomes and widely dissimilar mixes and scale of public and private services. Moreover, for many business firms the agglomeration economies derived from locating in a centre of given size may have little connection with its population level but instead may be related to the number of competing firms, the availability of specialised business services, and so on. Of course, for other firms (e.g. either those in consumer industry or those with specialised labour requirements) the population size of the city may be the chief determinant of market potential and labour market economies.

A further difficulty is that a city's population is an ambiguous measure depending very much on the arbitrariness of geographical and political boundaries. Intertemporal, cross-spatial and cross-cultural comparisons require standardisation in urban delimination. There are several solutions: aggregation of population within a given radius from the CBD; counting the population within a city region where the boundaries of the region are fixed by where the population falls to rural density levels; using a population potential index. Of course, the boundary problem arises with the adoption of other size variables too.

Even if, in the absence of output or income data, we can use population as a criterion for minimum threshold city size, is it really useful in determining an optimum? Assuming that data constraints require us to work with population, is the level of population more important than its spatial distribution? The focus of this Conference on national urban development strategies and the concept of a national system of cities presumably implies that within the national economy the spatial distribution of economic activity and population between cities is a key element in efficiency. Should not the argument be extended down the scale on the lines that the intracity spatial distribution of population is more important than the aggregate city population? Emphasis on the spatial distribution of population is justified by the fact that many external diseconomies are due to congestion and can be treated as a function of population density.[2] Of course, there is generally some relationship between the height of the density gradient and total city population so that city size and average population density tend to be closely correlated. Nevertheless, there are extremes within a given city size (e.g. Los Angeles and Chicago). There has been very little research into the relationship between urban costs and population density, but with city population held constant we might hypothesise a U-shaped function with heavy costs at high and very low average population densities (due to congestion and long average trip journeys respectively) and the bottom point of the function found at medium densities. This is a subject worthy of further research. If this hypothesis is confirmed, it suggests a switch of attention in discussions of optimality in the city (as opposed to the question of minimum thresholds) to efficiency in spatial structure rather than to the problem of aggregate size, and that analysis of optimality in the urban economy that abstracts from space is grossly inadequate.

Other criteria have been suggested for determining optimal city size. For instance, it is sometimes argued that urban development and infrastructure construction costs are lower in medium-sized towns which offer the economies of scale and accessibility to building material sources frequently not found in small centres yet do not suffer from the high costs of development in large cities. But the latter is primarily due to high land costs which as a transfer payment rather than a resource cost do not necessarily mean high real costs unless economising on land in selecting factor coefficients involves high costs, e.g. because of inelasticities in the supply of other.

Moreover, the discounted stream of urban development costs is such a small fraction of the total economic product of the city, and differences in streams among cities of different size much smaller still, that urban construction costs form much too narrow a criterion.

Another factor, stressed by Thompson (1965a and b), is the emergence of managerial diseconomies when the running of a city become a large-scale operation. In specific cases large cities may be run inefficiently because of a scarcity of high quality urban management personnel, but I doubt whether such diseconomies are inevitable. In the United Kingdom at least, impressionistic evidence points the other way: that small authorities are more likely to be inefficiently run primarily because successful urban management calls for highly skilled specialised staff that can be justified and afforded only by large authorities. Furthermore, to what extent are managerial diseconomies in business due to the physical impossibility of handling problems above a certain scale or to the inefficiencies of monopoly power? The latter scarcely apply in the urban context since the major cities of the system are in competition with each other to attract business firms and migrants. As Baumol (1967) pointed out, no modern city can let its tax rates or quality of urban services get too far out of line with its competitors. Moreover, within each city-size class efficiency is normally distributed among cities much as it is among firms of a given size in a particular industry. I suspect that the 'efficiency gap' between cities within a given size class may be about as wide as between cities of extremely different size.

Optimal city size discussions stress the importance of agglomeration economies, and an oft-quoted element in these is the benefit of a large, centralised labour market. Others at this Conference are much better qualified to speak on labour economics, so my remarks will be restricted to one or two obvious points. First, if there were a precise scale for an efficient city labour market, this could still mean substantial variation in city population sizes because of intercity differentials in labour participation rates. Second, the notion of an optimal scale is probably as dubious in regard to labour markets as to cities themselves. An efficient scale will vary according to the industrial composition of the area, the skills required, turnover rates in individual industries, and many other considerations. In general, however, the odds favour large labour markets, especially if they are spatially concentrated. They offer more chance of locational compatibility, i.e. that the aggregate demand for labour at the city's workplaces will match the total supply within travelling distance (see Goldner, 1955), and large size ensures threshold pool levels even for highly specialised types of labour.[3]

These arguments lead us to reject the value of the optimal city size concept, even in a simplified comparative statics framework. When we transfer the concept into a dynamic setting, its value depreciates even further. One reason for this is that urban growth dynamics require us to look at the individual city against the background of the wider system of which it forms a part, and in this wider framework it becomes clar that different cities perform different functions (purely local, regional, national, or cosmopolitan), and some cities perform all four. If we accept the implications of multifunctional cities, there is no possibility of a unique optimum since optimality makes sense only in the context of efficient achievement of objectives, and these objectives vary according to the type and functions of the individual city. Because of multiple functions and intercity specialisation, cities will differ in the local orientation of their industries and activities so that the local population threshold will be relevant in some cases but not in others. The efficient ranges of place size will be determined by the functions which each type of place has to perform. If the Hierarchical structure implied by differentiation in function extends to the national system of cities, there would appear to be a basic inconsistency between the optimal city size concept and an efficient national urban hierarchy. This conflict, though obvious, has been neglected in the literature.

If optimality were a relevant objective, a distinction would be need to be drawn between optimality in a single city and optimality in urban systems as a whole - the difference between partial equilibrium and general equilibrium. Thus, if city size equilibrium were an appropriate policy objective, attempts to pursue it would run up against the problems arising from the theory of second best since it is impossible to optimize the spatial distribution of cities instantaneously. It is perhaps possible to argue (Rashevsky, 1947) that a general interurban equilibrium is obtained when productivity per head is equalized. But it is doubtful whether the resulting distribution would be consistent with the national urban hierarchy. Moreover, whereas population might redistribute itself among cities in this way if neoclassical marginalist assumptions held, it is most unlikely that either interurban or rural-urban migration conforms to such a model. The benefits of higher urban productivity frequently fall upon businesses rather than the individuals, while the sharing of urban service costs through the tax structure and its incidence on different groups blurs the connection between population expansion and rising urban costs. Furthermore, the lack of urban data makes equalising per capita productivity infeasible as a policy goal. In any event an equilibrium based on equalising interurban productivity is static, wheras hierarchical distributions, such as the rank-size rule, represent a dynamic equilibrium in the sense that the distribution remains more or less the same overall even though the rank and rates of growth of individual cities in the hierarchy change (allometric growth).

Can the concept of optimal city size be made compatible with the existence of a national urban hierarchy? The answer to this question depends on what is meant by optimal city size. As we have seen, von Boventer's concept of subjective optima variable between firms in consistent with the hierarchy of city sizes. Second, instead of a single optimal city size it may be more meaningful to conceive of an optimal size for each rank order in the urban hierarchy. Third, if the looser concept of an efficient range of city sizes is interpreted very widely, a wide range may cover a large proportion of the city sizes in the hierarchy. Fourth, with increasing decentralisation from major cities into urban subcentres around the metropolises it becomes arbitrary whether we speak of one city or a family of linked cities. Since large cities of a prescribed size may imply conglomerations of smaller urban centres of variable size, it is possible depending on the definitions of urban areas adopted to envisage an optimal metropolitan area size which simultaneously embraces a hierarchy of city sizes. Finally, if optimal urban scale is measured not in terms of absolute size but of optimum density, the potential inconsistency may be resolved since average densities may vary less than absolute city sizes and the concept of optimum density (if it exists) is culturally and instutionally determined.

A minimum threshold size for a city may be more sensible than an optimal size. If a clear-cut minimum size could be found, its rationale could be based on two arguments. First, it could represent the population level needed to provide a demand for all major services without having to rely on other cities; once again, this raises acute questions of definition and industry/service mix. Second, it could imply a threshold in urban growth beyond which growth becomes self-sustaining. Theoretically, there is no reason why these two minima should coincide. Moreover, centres below the minimum might be quite efficient if conceived as part of a wider system, e.g. if they perform lower order functions, or were locations for key national industries with nuisance creating properties that require isolated sites. Criteria used to determine efficient minima for cities include demand thresholds, engineering cost functions for urban services, high growth rates (an analogy to the 'survivor' technique for estimating efficient firms), subjective observations of the range and quality of services offered in cities of different sizes, etc. The evidence is not wholly conclusive, but much of it supports the view that 200-250,000 is a minimum city population for providing a comprehensive range of services (Berry, 1968; Thompson, 1965 a; Cameron, 1970; Neutze, 1965; Clarke, 1945; Alonso, 1970b; Redcliffe-Maud, 1969; Hansen, 1970). This is not to deny that for many services a population of 30,000, 50,000 or 100,000 may be sufficient, while for certain specialised services (e.g. in the transport, medical and cultural fields) a population of at least one million may be necessary.

To sum up, to search for an optimal city size is almost as idle as the quest for the philospher's stone. Optimality may have meaning in the urban economy if it relates size to form and structure, but the crude measures of size in the literature have lacked any spatial dimension. It may be more meaningful to explore the concept of minimum critical size, though thresholds will vary according to the functions performed by the urban centre and the service mix selected by its residents. There may be more truth in the idea of an efficient range of sizes within each broad size-function class of cities. There may even be a theoretical maximum size beyond which cities are likely to encounter increasing obstacles to efficiency and growth. But these modifications destroy the concept of a unique optimum. This is, fact, reassuring for the dynamic analysis of city systems since these invariably seem to take the form of a hierarchical structure.

III. THE RANK SIZE RULE

There has been little by way of formal analysis on the distribution of city sizes apart from the classical concepts of the rank size rule and the urban hierarchies derived from the application of Christaller-Losch market are principles. It can be shown that there are certain similarities between rank-size and central place principles and that the two distributions can be made consistent with each other (Parr, 1970). However, there are basic differences First, the rank-size rule is an empirical relationship ('a fact in search of a theory'?) while central place hierarchies can be deduced by a priori reasoning. Second, although this is by no means a necessary implication, central place hierarchies tend to refer to the regional system while the rank size rule is usually applied to nations.

We do not need to give much attention to the statistical aspects of the rank size rule, for these are familier. The rank-size rule states that we can derivie the rank in the urban system of a centre for knowledge of its population size:

$$r_x = aP_x^{-q} \qquad (1)$$

where P_x = population of centre x, r = rank, and a and q are constants.[1]

A special case of (1) is

$$P_r = \frac{P_1}{r} \qquad (5)$$

in which a = P_1, the population of the primate city, and q = 1. Where the special case is valid, we obtain the population of any centre by dividing the population of the leading city by the rank of the centre in question.[2] Finally if the entire population of the system is urban then the total population,

$$P_t = a\Sigma r^{-q} \qquad (3)$$

Weiss (1961) demonstrated that, by choosing appropriate values for the parameters a and q, it is possible to fit the cumulative population data for the U.S. to within 1 per cent, and that this covered 4,000 cities and towns and 59 percent of the population. If the sum held until all the U.S. population was exhausted, 12,000 centres would be required.

There are at least three important questions raised by the rank-size rule. How far does it apply? Can it be explained? Does it represent, in any sense, an optimal distribution of city sizes? The empirical evidence for the rank-size rule is mixed: it holds in some countries but not in others; in one or two cases, perhaps Western Europe is the most notable example, it fits a multinational system better than for a single nation. If anything, it is more likely to apply in large countries with a long history of urbanisation and with a complex economic structure. At the same time, Berry (1961) found no clear relationship between conformity to the rank size and the level of economic development; though others (Bell, 1962; Friedmann, 1963; and Boal and Johnson, 1965) disagree.

A more important question, once we accept that in many highly urbanised economies there is a distribution of city sizes which approximates, to a greater or lesser extent, to a rank-size relationship, is whether there is a satisfactory explanation of why systems of cities structure themselves in this way. There is no acceptable economic theory but there is wide support for an explanation that is partly based on statistical theory, partly on an analogy drawn from thermodynamics (Vining, 1965; Berry, 1961 and 1964; Curry, 1964 and 1967; Olsson, 1966). Vining, for example, argued that the rank-size rule is a time dependent stochastic process and that the distribution of city sizes maintains stability (a form of 'statistical equilibrium') even though there is a great deal of individual flux, e.g. city populations changing over time, rises and falls in rank, births and deaths of towns, etc. Olsson demonstrated that the most probabl or equilibrium state of a system of settlements is equivalent to the rank-size rule, and that 'a system of cities obeying the rank-size rule is in a state of equilibrium in which "entropy" has been maximised'.[6] Curry (1964) similarly showed that the rank-size rule conformed to the most probable distributin of a random arrangement of a given number of people in a given number of cities. Randomness has the effect of avoiding too much concentration in large cities and ensuring that the population is spread (though not evenly of course) among all cities.

Even if the rank-size rule is the most probable state of the system of cities, this does not mean it is in any way an optimal distribution. If it represents an entropy-maximising state there may be tendency for economists by force of habit to equate the equilibrium solution whith the desired or optimal solution, but this is a danger to be resisted. Indeed, there is a degree of incompatibility in the notion of an optimum resulting from a process that is essentially stochastic. On the other hand, it is consistent with the mass of individuals making optimal location decisions since probability theory suggest that the total effect of large number of optimising decisions may appear as random. But as we know, once we acknowledge the existence of imperfections in competition and externalities, the sum of individual optimising decisions is not necessarily a societal optimum. It would be foolish to treat the rank-size distribution as an ideal state and to adopt an urban policy aimed at correcting deviations from it. What the evidence on distribution of city sizes does suggest, however, is that a strongly hierarchical distribution always results from the operation of competitive forces in an urbanised economy. This implies that a national urban hierarchy is perhaps an efficient distribution for achieving national and regional objectives.

IV. THE NEED FOR A NATIONAL URBAN HIERARCHY

It is well known that a hierarchy of urban centres is the efficient way of organising production and distribution within a region, i.e. the central place system. Similarly, the national urban hierarchy performs a number of important functions in the national economy which may make it an efficient medium for the distribution of the total population even if some individual cities in the size distribution are outside the efficient size range from the point of view of the cities when considered alone.

Although central place theory, with its emphasis upon supplying central services to the hinterlands around towns, can be used only to explain the regional hierarchy, and then convincingly soley in rural regions, it nevertheless has relevant implications for analysis of the national hierarchy.[7] First, it accounts for the lower orders in the distribution of city sizes since the national hierarchy is merely the sum of all regional hierarchies. The small sub-regional urban centres will feature as low ranking towns in the national system, leaving to be explained in other ways only the system formed by the leading cities in each region and why these assume a hierarchical structure from the standpoint of the national economy. Second, the principle behind central place theory - that industries have market areas of different size - is still of value at the national level.

The modifications to the Christaller central place analysis needed for explanation of the national hierarchy owe a lot to Lösch (1954), Stolper (1955), Tinbergen (1961) and Bos (1965). The basic difference is that whereas in the central place hierarchy centres of the same size always produce the same order of goods and services and the industries with large market areas and scale economies are always produced in the larger cities, this is not necessarily the case with the national distribution. City size is positively linked with the number of activities located there and since economies of scale and market area threshold differentials rule out the presence of each industry in each city, we obtain cities of different size. The largest city will be that which captures the highest proportion of national (and international) market industries while at the same time encompassing most other industries with smaller market areas. The place of other cities in the hierarchy will similarly be determined by how many particular industries and of what size it attracts. This cannot be predicted with accuracy since the attractive power of cities depends upon the quality of their services, the size and compactness of their markets, and the strength of their agglomeration economies.

The national hierarchy lacks the systematic order and the predetermined pattern of the central place regional hierarchy. Towns of similar size may produce quite different 'mixes' of goods and services, the same industry may locate in cities of different size, and even large scale economy industries may be found in small cities. On the other hand, the larger a city the more likely that it will be specialised in national market industries.

The national hierarchy of leading cities fulfils other purposes that give justification to the view that it is an instrument for achieving national growth. First, the urban hierarchy is an efficient vehicle for transmitting new technology, managerial expertise and general economic functions from the centre of the economy to the periphery. This permits social and economic change to 'leapfrog' over distance and avoid the slower gradual diffusion over space from the central city. This transmission function is aided by the fact that many modern forms of business organisation (in commerce, finance and industry) are themselves hierarchical with head offices and centres of decision making in the metropolitan centres and their decision trees spread out spatially down the urban hierarchy. Given the large scale and multiple establishments characteristic of modern business, a hierarchy of cities make it easier to distribute the hierarchical structure and functions of business organisation over the economy as a whole.

Second, a hierarchy of cities permits specialisation, division of labour and differentiation in economic function. Market size requirements, infrastructure needs and agglomeration economies differ between firms in the same industry (many industries have a size distribution of firms similar in form to the distribution of city sizes) and between industries. A hierarchy of cities offers firms a wider choice in location and enables them to operate more efficiently (von Boventer, 197). Third, the hierarchical structure of cities dominant within their own regions enables each city to function in a manner appropriate to the size and character of its hinterland region. The leading city in a peripheral region, for instance, clearly plays an important role in the development of that region (the growth centre strategy). Since regions vary in area, population and level of economic development it is inevitable that their leading cities will also vary in size.

V. THE SPACING OF CITIES

A distinction must be drawn between the size and the spatial distributions of cities in the national economy. The urban size hierarchy and the urban spatial hierarchy are not necessarily symmetrical. For instance, the size hierarchy may appear very efficient in terms of the distribution of population among existing centres but it could be grossly sub-optimal because of irregular spacing.

Geographers and other spatial analysts have developed one or two simple rules to determine the average distances between centres. One of these (Stewart, 1958) is that the distance between two cities is proportional to the multiple of their populations, i.e. $d_{ij} = P_i P_j / G$, where G is a constant. Another (Curry, 1967) is that in an n order hierarchy the distances between centres in the same rank can be related to distances between centres of other ranks by the function $d_n = d_{n-1}^{1.5}$. This is a special case of the Christaller-Lösch general principle that $d_n = d_{n-1}\sqrt{k}$, where k = number of settlements supplied by n of rank n-1. Thus, the results vary according to whether the nesting pattern refers to 3, 4 or 7 centres.

While it is true that economic factors lie behind these principles in that the distances can be deduced from market area analysis and while limited conclusions may be drawn from them (e.g. the obvious point that average distance between centres increases as we ascend the urban hierarchy), we should not expect too much similarily between the theoretical and actual spatial distributions. There are three main reasons for this. First, the distances referred to are average not uniquely determined distances, bound by the assumptions of the market area models (e.g. uniform population densities). In particular, the dispersion around the theoretical average increases with city size because large cities (and hence the number of observations) are few.

Second, the existing distribution of cities in a developed economy reflects historical patterns of growth, and the initial location of many cities may have been due to factors (raw material orientation, access to strategic transport routes, random elements) other than market area influences. For instance, arguing ab initio in a uniform plain model, the main centre should be located so as to minimise transport costs to and from all parts of the economy, i.e. at the centre of the plain. Yet in many countries, particularly outside Europe, the larges metropolis is often found on or near the coast reflecting the fact that the countries were originally opened up for development by colonisation from overseas. The international and entrepôt functions of these leading cities may persist almost indefinitely. Thus, the historically evolved pattern of urban settlement constrains subsequent development. For the purposes of analysis, the fact that the location of some cities is predetermined (due to spatial irregularities and good natural sites as well as to historical factors) make it much simpler to assign locations to the others, but it does distort the actual from the theoretically predicted spatial distribution. Third, the distance between cities is not independent of the size of the economy, particularly with regard to area and population density. For instance, Lösch found that in the Unite States rural densities declined with movement westward and average interurban distance increased. Even more obvious, interurban distances will tend to be shorter in small and/or highly urbanised countries.

A more sophisticated economic analysis of interrelationships in space between urban centres has been presented by von Boventer (1969 and 1970). He argues that two main factors influence the ideal location for a city or a town: agglomeration economies gained by locating near a larger centre and hinterland effects which are small (or negative) near large cities because of their competitiv power but are strong at a distance because of the protection of a sheltered market area. The optimal distance for a city from the nearest larger or equal sized city is that which maximises the sum of agglomeration and hinterland effects. The ideal distance may be difficult to pinpoint, but von Boventer argues that there is some intermediate distance (the pessimum distance) at which the city will be worst off. Thus, an urban centre has better growth prospects if it is either close by a vigorous bigger city or far away from all competitors.

The broad policy implications of this analysis are more important than any attempt to determine precisely either the ideal or the pessimum distance.[8] From the point of view of maximising growth in the economy as a whole, it is undesirable that large cities should be close together since a large city can create its own agglomeration economies which can support smaller centres and (via a growth centre strategy) regional expansion at a distance. Small urban centres, on the other hand, have few positive hinterland effects and a location far away from a metropolis offers no benefit. Thus, small centres thrive best if they are not too far away from a metropolitan centre so that they can profit from agglomeration economies and spillover effects, though it is obviously an advantage if there are no competitors of similar size nearby. This argument is strengthened if there are diseconomies of scale in urban size, since the small city near the large metropolis reaps the benefits of agglomeration without the pains of large size. In Alonso's (1970b) succinct phrase: 'Policies of small and far, which are not uncommon, perhaps should be small and near, and big and far'. Furthermore, by adopting a development axis approach with centres built up along axes from or between metropolitan cities it may be possible to retain access to agglomeration economies without having to sacrifice hinterland effects. The whole argument suggests that a polinucleated megapolitan structure, far from being an untidy and ugly urbanised sore, could be an efficient constellation of urban centres for fostering national growth. The most appropriate method of stimulating growth is lagging regions, on the other hand, could be to reinforce their largest cities.

VI THE GOALS OF A NATIONAL URBAN POLICY

Even though several policies pursued by governments in developed economies have had the effect, and sometimes the explicit intention, of influencing the spatial distribution of population between cities and towns, it would be misleading to speak as if these countries had a clear national urban policy. In the UK for instance, measures have been implemented in an ad hoc and piecemeal fashion without any conception of a comprehensive strategy. Also, since they were introduced very late in the day in a heavily industrialised, highly urbanised society, such measures have not been able to plan the development of an urban system (as in Turkey and much less successfully in Brazil) but merely to modify an old-established structure, and in a very minor way. British 'policy' has had three major aspects: decentralisation of people and activities from the great metropolises (primarily London but also Liverpool, Manchester, Birmingham, Tyneside and Glasgow), the main manifestation of which has been overspill policy; the related, but overrated, New Towns policy; and more recently, examination of the possibility of stimulating growth centres in the Development Areas, but little action. It is unnecessary to detail the evaluation and execution of these policies. They are well known, and in any event they do not add up to a national urban strategy.[9] The New Towns, despite being the most distinctive feature of UK urban policy, served predominantly especially in the early years as overspill pockets for London, and did not represent the emergence of a strategy for a national system of cities. From this point of view, they make less sense than the metropoles d'equilibre of France which have the dual function of acting as a counterweight to Paris and of promoting the development of their hinterland regions.[10] To the extent that the new towns were an innovation, this was as social and architectural experiments within the towns not as a major influence on the urban hierarchy, since they merely added another score or so of towns to the 900 or so smaller towns of less than 75,000 in Great Britain.

It is possible for a national urban policy to be based on the pursuit of a single objective. Alternative objectives might be:

(i) minimisation of interurban migration;
(ii) prevention of urban imbalance;
(iii) convergence of interregional income and growth differentials[11];
(iv) maximisation of innovation diffusion potential;
(v) creation of a market for cities;
(vi) equating the private and social costs of urban expansion.

Many of these are too broad for implementation and need to be translated into a more specific form for the guidance of policymakers. Some are ambiguous. Others are clear but are either too costly or impossible to make operational. The points may be illustrated by commenting on each goal.

In a country without a significant rural surplus of labour, goal (i) implies that each city grows at its natural rate. Although this could alter the city size distribution if there were large interurban differences in fertility, this is unlikely to bring about much change in urban rankings.

However, despite locational preferences for a familiar environment, such a goal is unattainable. It may make more sense to modify the objective to minimising 'involuntary' migration, but this raises acute questions of definition. In any event, any policy involving a reduction in the overall rate of migration is likely to interfere with national growth objectives.

Objective (ii) is not very specific, though it clearly implies restrictions on the growth of large metropolises. However, we have seen that there is no evidence to suggest that a more even size distribution of large cities implies greater efficiency and higher welfare than a skewed distribution. Even if an even distribution were considered desirable, it is doubtful whether available policy instruments could achieve much by way of cutting down the size of great cities. Moreover, there are political difficulties in the way of strengthening these instruments. Large city governments are powerful, and are unlikely to take kindly to national policymakers who attempt to reduce their territory and tax base. The interference in local autonomy that this national-local conflict necessarily involves is itself a far from negligible social cost.

The third goal is also difficult as long experience with regional policy shows. In regard to urban policy it requires controlling city growth in rich regions and expanding urban centres in lagging regions. If the rich regions contain the largest cities, then operationally this policy becomes similar to the urban growth strategies pursued in most countries. However, as it stands it may be an oversimplification since the argument assumes that income and growth indicators move together. If large cities have high proportions of low income residents the policy implications may be rather different with a possibility of conflict between growth and income equalisation (equity), i.e. high growth rates may require large cities in backward regions but these may create rather than reduce poverty.

The maximisation of innovation diffusion potential, though important, is unlikely to prevail as a single urban policy goal but to be associated with others. We know that realisation of such a goal requires a national hierarchy of cities with strong communication links between them and multi-plant corporations operating in or near many centres in the hierarchy. Where our knowledge is deficient is in identifying more precisely the most efficient hierarchical distribution for the diffusion of innovation, technical progress, new management methods, etc. Accordingly, other than investing in the communications networks it is difficult to know what pursuit of this goal implies by way of policy measures.

The argument of goal (v) is that community welfare is increased by offering households and firms a wider freedom of choice in the types of urban centre in which they choose to locate. Pursuit of this goal simply requires the government to stimulate product differentiation in the types and quality of urban living by developing a system of cities of widely different size and with very varied environments and urban life styles. This strategy, the creation of a 'market for cities' is an extension on a wider front of Tiebout's (1956) case for a varied interurban mix of urban service-tax structures. It probably requires extensive Government subsidies since more product differentiation needs urban experimentation which involves risks and doubtful payoffs. A serious objection to the relevance of this goal is that an extremely wide choice in urban environments, life styles and service-tax mixes can be found _within_ a very large metropolitan area without having to create a differentiated system of separate cities.

Goal (vi) is a favoured single objective. It requires measures to redistribute interurban population until the private and social costs of urban expansion are matched in each city. The rationale is that the divergence between social and private costs leads to excessive growth of cities. This happens because households and firms make location decisions on the basis of average costs, but the marginal costs of their entry into a city fall on the entire community, and these costs are much higher than average costs because cost curves are rising. Thus, market decisions result in cities becoming too big unless immigrants can be forced to pay the marginal costs associated with their arrival. The trouble with this objective is that it is incomplete and it is non-operational. It is incomplete because it ignores the benefit side of the picture. In-migrants may also create social benefits and these may or may not exceed the gap between social and private costs. It is non-operational because we cannot accurately measure social costs and benefits and hence cannot fix the appropriate levels of taxes (subsidies) even if a marginal cost pricing policy was politically acceptable. Moreover, there are dangers in applying a static pricing framework to a dynamic problem. For instance, rising marginal costs of in-migration might be incurred only in the short run because the rate of in-migration is temporarily too fast to be absorbed. In other words, they may result from inelasticities in short-run supply functions rather than from higher social costs in the long run.

These comments suggest that the apparent simplicity of a single goal national urban policy is deceptive. The overall comprehensive goal is often vague and frequently subsumes several other implicit and not necessarily consistent objectives. It is usually not specific enough for us to translate it into clear criteria for action. Many of the dimensions of the goal cannot be quantified, and this means that optimising decisions must be to some extent subjective since they reply on the policymaker's own weights. It follows that a workable strategy for a national urban policy is likely to be based on several goals with all the possiblities of conflict that multiple goals imply. The nature of these goals will vary from one society to another according to levels of economic development and socio-cultural values, as will the priorities among selected goals. In most developed economies, however, we would expect to find the following minimum set:

(a) the pursuit of national growth
(b) the promotion of interregional equity
(c) a 'quality of life' goal.

There is, of course, the possiblity of other goals, either independent from those on the list or subsumed within them.[12]

Goal (a) has implications for the distribution of city sizes, both in regard to the structure and spacing of the hierarchy as a whole and to the importance of leading cities as generators of growth and innovation. Although goal (b) may call for controls on urban growth in prosperous regions, mainly via redirecting expanding industry to lagging regions, its main focus will be on building up smaller urban centres in backward regions to that critical minimimum size which guarantees agglomeration economies. The 'quality of life' goal -goal (c)- is possibly the most important component of an urban policy strategy, yet it is the most nebulous. It refers primarily to the possible divergence between productivity and welfare in cities due to non-measurable social costs and intangible benefits. We still know very little about what constitutes a good environment for urban living, though we do know that perception of and response to environment differ a great deal from individual to individual and between groups. Moreover, the social costs, however defined, may be offset by higher real incomes and less quantifiable benefits associated with large city life; how far this is true depends upon whether individuals are 'forced' to live in large cities or whether their residential location decisions reflect a 'free choice' and a personal favourable balance of benefits over costs. Furthermore, if higher social costs are generated in large cities, it is unclear whether this is a direct consequence of size itself or simply reflects remediable managerial and organisational inefficiencies, the absence of pollution control policies, or lags in the response of urban governments to past population pressure. If the former there may be a prima facie case for decentralising population into smaller urban centres, but if the latter factors explain the high social costs the answer may lie in more effective city management and improvements in intracity spatial distribution.

The presence of multiple goals and the fact that priorities will differ among national, regional and urban policymakers mean that it is difficult to prescribe an optimal strategy to fulfil the stated objectives. Goal conflicts will have to be reconciled, the 'best' technical solutions will have to be modified by the constraints of political feasibility and by the need to satisfy community and individual preferences. Optimisation is scarcely practicable. In particular, it is impossible for all the objective of urban policy (faster growth, the revival of lagging regions, the control of environmental nuisances, etc.) to be satisfied by a unique optimum in the distribution of cities. For instance, even if it were true that large cities were always more efficient economically than small, this would not rule out the possibility that a hierarchy of sub-optimal cities could be more 'efficient' in relation to goal achievement (including growth goals) than the encouragement of a pattern of a few ever-increasing metropolises. Finally, in a multiple goal framework it is difficult to know whether a particular policy measure works in the right direction since its impact on the attainment of one goal may be offset by unforeseen disturbances on the attainment of others. All this amounts to a plea for very careful pragmatism rather than ambitious and inevitably abortive attempts to optimise the size and spatial distribution of cities.

VII THE INSTRUMENTS OF URBAN POLICY

Even if urban policy goals have been precisely specified and it is known what changes will achieve them, there remains the problem to whether the instruments are, or can be made, available to bring about the required changes. It is arguable that a centrally planned economy would be needed to attain a particular distribution of city sizes in a spatial framework, and even in planned economies experience in attempting to halt the growth of large cities has not been very successful. Secondly, and this is the crucial point, the instruments available within the city to change the interurban spatial distribution of activities are more powerful and their effects more predictable than the instruments available for modifying the interurban spatial distribution. This is primarily because planning, zoning and land-use controls are more effective (though not necessarily more efficient) than investment incentives, tax-subsidy measures, etc. Moreover, measures to improve the efficiency and organisation of the great metropolises may not only be more practicable than measures to prevent their growth but also more important in the sense that they will affect more people.[13] The arguments for emphasis on metropolitan rather than city size distribution solutions to policy problems should not be interpreted as a case for leaving policy in the hands of local governments. For if polinucleated clusters of cities are desirable and efficient (because they create large markets and offer a flexible structure for expansion), they need intermetropolitan and national policies to cope with common problems - transport, water and sewage, open space and recreation requirements, and air pollution.

In distinguishing between policy instruments, we must be careful to draw the line between those directly and those incidentally related to urban policy.

For example, the central government's national policies - industrial policies, monetary and fiscal measures, transport policy, health, education and welfare expenditures - have differential impacts on cities. Regional measures occupy an intermediate role since policies to redistribute industry between regions are bound to have large urban side-effects because most population and economic activity is urban. Measures to influence the interregional distribution of industry (investment incentives, tax allowances, payroll subsidies, prohibitive controls such as i.d.c.s.) can also be used to implement urban policy goals, but only in rare circumstances will their use in this context be compatible with their role as a regional policy instrument. For instance, controls on office building in Greater London since 1964 have been primarily associated with attempts to persuade firms to relocate in the Outer Metropolitan Area, and this had done nothing to correct imbalance in expanding service industries between the South East and the outer regions of Britain. Other measures, such as the provision of infrastructure and social overhead capital, simultaneously serve national, regional and urban ends. In this sense, decisions on the location of infrastructure might be regarded as the unifying theme in the co-ordination of national urban policy, except for the facts that infrastructure decisions are taken at different levels of government and that successful realisation of urban goals also requires a sequence of compatible private decisions by households and firms to rationalise and justify the prior (or associated) infrastructure decisions.

Apart from public infrastructure investment, the two most obvious urban policy instruments are (i) planning, land and zoning controls and (ii) pricing methods, such as congestion taxes, user charges for public services, etc. Both can be justified in general by reference to 'externality' arguments. The trouble arises when attempts are made to justify the application of specific measures in relation to the achievement of urban policy goals, especially if these goals include an efficiency in resource allocation component. The repercussionary effects of land use controls are difficult to trace, while the importance of non-measurable social costs makes it impossible to set 'optimal' congestion taxes. In addition, the widespread use of pricing methods to control urban scale is suspect on the grounds of political feasibility. The crudity of our policy instruments once more reinforces the case for pragmatism. In devising an urban growth strategy, goals set as targets are more likely to be attainable than goals expressed in terms of optimal efficiency. The best we can do is to take action in response to specific and clear-cut problems in the hope that we nudge the spatial allocation of resources a little in the desired direction.

VIII CONCLUSIONS

If it implies an optimal distribution of national population through the nation's existing and planned urban centres, a national system of cities is not a sensible object of public policy. This is not to say that a national urban policy is undesirable or unnecessary. But such a policy should be goal-oriented, and we can identify several separate if overlapping and frequently conflicting goals. Moreover, the magnitudes of change needed to achieve these goals are in some cases not easily, if at all, measurable. With many goals, and when some of these cannot be translated into quantifiable objectives, the case for optimisation loses much of its strength. Certainly there is no unique spatial distribution of population which can be said to achieve these goals. Von Boventer (1970) has pointed out that deriving an optimal distribution of population is difficult. I will go further; in a policy context, it is impossible.

A hierarchy of cities is an efficient system for promoting national growth and for producing and distributing goods and services to society. This suggests that it would be foolish to attempt to equalise the size of cities, but it does not help us to decide whether one hierarchical structure is superior to another. There are a few general indications. First, variety in urban form, structure and environmnent both in regard to differentials in city size and within a specific size-class is probably a good thing because it offers individual households and firms a wider choice, Given heterogeneous tastes, wider choice implies greater welfare. Second, if a first-order metropolis, is a seedbed for innovation, managerial expertise and growth and a 'port of entry' for new technology and ideas into a region, then every region should contain a large city (relative to the size of the region). If this condition is fulfilled, a national hierarchy of regional cities will further devolve into sets of central place systems. Third, whatever the size distribution of the top level of the national hierarchy, its efficiency depends as much on the quality of the transport and communications networks linking the major cities as on the balance between agglomeration economies and urban costs within them.

The promotion of growth centres in lagging regions is in many cases a valid policy objective, but it is much more relevant to regional than to urban policy. A growth centre strategy involves an attempt to maximise regional growth potential via spatial concentration of development and to obtain cost-effectiveness in urban infrastructure spending. Much more central to a national urban policy is what, if anything, should be done about a nation's large cities. We know that all is not well, that our great metropolises have generated considerable social costs. The fact that many of these are difficult, or impossible, to measure cannot deny their existence and importance. What we do not know, however, is how far these social costs may be offset by social benefits in the form of higher urban productivity. Some, but not all, of this higher productivity may be captured in higher real incomes, but the incompleteness of urban data makes it difficult even to quantify the measurable component. If higher costs were outweighed by higher productivity, we would have not a diseconomy of scale question but an income distribution question, since the burden of social costs falls heavily on those too poor to evade them, while the benefits of higher urban productivity are not equally shared but accrue to owners of urban land, business corporations and other monopoly groups. Furthermore, it is unclear whether higher social costs are inherent in the phenomenon of large size or whether they might not be transient due, say, to a too fast rate of in-migration relative to the increase in urban capacity i.e. a problem of absorption rate rather than absolute scale. If so, the appropriate remedies would be to monitor and control the rate of expansion of large cities not to attempt to reduce them in size.

Even if we suspect that the net social benefits of the present distribution of urban population are negative, there are two further considerations. First, the policy instruments available in a mixed economy are probably not strong enough to reduce the size of large cities. By influencing the distribution of jobs and the provision of housing and infrastructure, however, it may be possible to alter the spatial distribution of urban population in a period of population expansion by operating on the relative growth rates of individual cities. Second, a danger in pursuing, possibly ineffective, measures to persuade people and activities to leave the large metropolises is that this may provide a pretext for neglecting the pressing problems of improving the urban environment and central city poverty. These problems are not directly related to scale, and will persist even if the size of large cities could be reduced.

Does all this mean that nothing is left of an urban policy strategy? Not at all. However, it implies that such a strategy should be pragmatic, even piecemeal, rather than based on an optimisation approach aiming to achieve a dynamic spatial equilibrium at one stroke. Moreover, a national urban policy ought not to include measures to tain optimal size for an individual city. Optimality in this context has no real meaning. Even if this were not the case, from the viewpoint of national policy an artificial system of cities of 'optimal size' would be less efficient than a hierarchy both for economic growth and for providing an array of different environments for businesses and people. Furthermore, there is not a single unique efficient distribution of city sizes, but many, and given multiple goals we should not be too worried about minor irregularities in the city size distribution.

A pragmatic approach does not rule out measures that affect the national urban hierarchy. These might include action to boost leading cities in backward regions or to bring small cities up to a minimum critical size compatible with efficiency and continued growth. Also, measures to decentralise activities from our largest cities are not inconsistent with a national urban policy. But such measures are unlikely to be successful if they aim at reducing the scale of our metropolitan areas by redistributing population to other distant urban centres in other regions. Better prospects are offered by dispersing population from the central city itself to interconnected smaller centres within the metropolitan region. The promotion of a polinucleated system enables the smaller cities to benefit from the agglomeration economies offered by access to the metropolis. It is also more likely to achieve results since it probably satisfies locational preferences much more than an interregional interurban redistribution.

Most important of all, however is that the most effective policy, especially if we accept the argument that there are many efficient city size distributions, could well be to improve the efficiency and management of our large cities by acting on the intra-city spatial distribution and by correcting the most obvious resource misallocations within the city. The large city is an important engine for growth, and we know that rapid urban expansion may improve welfare rather less than it boosts growth. Is it not, therefore, a sensible urban policy objective to concentrate on improving the urban environment and 'livability' within the city? In this way more people will benefit than by measures to stimulate out-migration from the cities which might adversely affect growth potential without achieving a commensurate reduction in social costs.

REFERENCES

W. Alonso, 'What are New Towns For?' Urban Studies, 7 (1970a), 37-55.

W. Alonso, 'The Question of Urban Size', Paper read at Regional Science Association Conference, London, (August, 1970b).

W.J. Baumol, 'Macroeconomics of Unbalanced Growth: the Anatomy of Urban Crisis', American Economic Review, 57 (1967),pp. 415-26.

M.J. Beckmann, 'City Hierarchies and the Distribution of City Size', Economic Development and Culture Change, 6 (1958), 243-8.

M.J. Beckmann, Location Theory, (Random House, 1968).

G. Bell, 'Change in City Size Distribution in Israel', Ekistics, Vol. 13 (1962).

B.J.L. Berry, 'City Size Distribution and Economic Development', E.D.C.C., 9 (1961), 573-87.

B.J.L. Berry, Geography of Market Centres and Retail Distribution, (Prentice Hall, 1967).

B.J.L. Berry, 'A Summary - Spatial Organisation and Levels of Welfare: Degree of Metropolitan Labour Market Participation as a Variable in Economic Development', Research Review (EDA), July 1968, pp. 1-6.

F.W. Boal and D.B. Johnson, 'The Rank-Size Curve: A Diagnostic Tool?'. Professional Geographer, Vol 17, 21-3 (1965).

H.C. Bos, Spatial Dispersion of Economic Activity, (North Holland, 1965).

G.C. Cameron, 'Growth Areas, Growth Centres and Regional Conversion', Scottish Journal of Political Economy, 21, (1970), 19-38.

C. Clark, 'The Economic Functions of a City in Relation to its Size', Econometrica, 13 (1945), 97-113.

L. Curry, 'Central Places in the Random Spatial Econòmy', Journal of Regional Science, 7, (1967), 217-38.

O.D. Duncan, 'The Optimum Size of Cities', in J.J. Spengler and O.D. Duncan (eds), Demographic Analysis, (Free Press,1956).

J. Friedmann, 'Economic Growth and Urban Structure in Venezuela', Guademos de la Sociedad Venezolana de Planificacion, Special Issue.

A. Girard and H. Bastide, 'Les Problemes demographiques devant l'opinion'. Population, 15 (1960).

W. Goldner, 'Spatial and Locational Aspects of Metropolitan Labour Markets', American Economic Review, 45 (1955).

S.P. Gupta and J.P. Hutton, 'Economies of Scale in Local Government Services', R.C. on Local Government in England, Research Studies, No. 3 (1968).

N.M. Hansen, 'A Growth Centre Strategy for the United States', Review of Regional Studies, 1 (1970), 161-73.

R. Higgs, 'Central Place Theory and Regional Urban Hierarchies: An Empirical Note', JRS, 10 (1970), 253-5

W.Z. Hirsch, 'Expenditure Implications of Metropolitan Growth and Co-ordination', Review of Economics and Statistics, Vol. 41, 1959, pp. 232-41.

E.M. Hoover, 'The Concept of a System of Cities: A Comment on Rutledge Vining's Paper', 196-8, E.D.C.C., 3 (1955)

E.M. Hoover, 'Transport Costs and the Spacing of Central Places', pp. 255-74, P.P.R.S.A., 25, (1970).

W. Isard, Methods of Regional Analysis, (MIT Press, 1960).

K.S. Lomax, 'Expediture per Head and Size of Population'. Journal of the Royal Statistical Society, 106 (1943).

A. Lösch, The Economics of Location, (Yale U.P., 1954).

C.H. Madden, 'Some Spatial Aspects of Urban Growth in the U.S.', E.D.C.C., 6 (1958-9), 371-87.

F.T. Moore, 'A Note on City Size Distributions', E.D.C.C., 7 (1959), 465-6.

G.M. Neutze, Economic Policy and the Size of Cities, (Kelley, 1965).

G. Olsson, 'Central Place Systems, Spatial Interaction and Stochastic Processes', P.P.R.S.A., 18 (1966), 13-45.

J.B. Parr, 'City Hierarchies and the Distribution of City Size', J.R.S., 9 (1969), 239-54.

J.B. Parr, 'Models of City Size in an Urban System', pp. 221-53, P.P.R.S.A. 25, (1970).

N. Rashevsky, Mathematical Theory of Human Relations, (Bloomington: Principia Press, 1947).

Redcliffe-Maud, Lord, Royal Commission on Local Government in England, (Cmnd. 4040, 1969, H.M.S.O.)

L. Rodwin, Nations and Cities: A Comparison of Strategies for Urban Growth, (Houghton-Mifflin, 1970).

J. Rothenberg, 'The Economics of Congestion and Pollution: An Integrated View', American Economic Review, 60 papers (1970), 114-21.

Royal Commission on Local Government in Greater London, Cmnd. 1164 (1960).

T.M. Stanback, Jr. and R.V. Knight, The Metropolitan Economy: The Process of Employment Expansion, (Columbia U.P., 1970).

C.T. Stewart, Jr., 'The Size and Spacing of Cities', in H. Mayer and C.F. Kohn, Readings in Urban Geography, (Chicago U.P., 1959).

South East Joint Planning Team, Strategic Plan for the South East (HMSO, 1970).

South East Economic Planning Council, The Strategy for the South East, (HMSO, 1967).

W.R. Thompson, A Preface to Urban Economics, (John Hopkins, 1965a).

W.R. Thompson, 'Urban Economic Growth and Development in a National System of Cities', 431-90, in P.M. Hauser and L.F. Schnore, The Study of Urbanisation (Wiley, 1965b).

C.M. Tiebout, 'A Pure Theory of Local Expenditures', Journal of Political Economy, 64 (1956), pp. 416-24.

J. Tinbergen, 'The Spatial Dispersion of Production: a Hypothesis', Schweizerishe Zeitschrift für Volkwirtschaft und Statistik, 97 (1961).

R. Vining, 'A Description of Certain Spatial Aspects of an Economic System', 147-95, E.D.C.C., 3 (1955).

E. von Böventer, 'Derterminants of Migration into West German Cities, 1956-61, 1961-6', P.P.R.S.A., 23 (1969), 53-62.

E. von Böventer, 'Optimal Spatial Structure and Regional Development', Kyklos, 23 (1970), 903-24.

H.K. Weiss, 'The Distribution of Population and an Application to a Servicing Problem', Operations Research, 9 (1961), 860-74.

FOOTNOTES

[1]Isard's objections to weighting procedures in aggregating separate cost curves remain as valid as ever (Isard, 1960, pp. 527-33).

[2]Rothenberg (1969) has shown the close links between pollution, possibly the most important externality, and congestion.

[3]A recent United States study (Stanback and Knight, 1970) has shown that in the 1950s it was the labour markets in the 200,000-1,600,000 range that grew fastest.

[4]Thus $\log r_x = \log a - q \log P_x$
so that if we plot rank against size on double log paper we should, if the rule holds precisely, obtain a straight line with the slope of -q.

[5]The special case is very much a special case. The leading metropolis in the system may diverge widely from the city size needed to support equation (2) while Moore (1959) showed that in both the U.S. and the U.S.S.R., $q<1$ and declines over time.

[6]The concept of entropy is derived from the second law of thermodynamics. This specified the direction in which a closed system in disequilibrium will move in order to reach equilibrium: thus, energy moves from hotter to cooler bodies. Entropy is a measure of the degree of equalisation reached within a system, and entropy is maximised when the system is in equilibrium.

[7]By this, I mean economic implications. It has been argued (Hoover, 1955; Beckmann, 1958; Parr, 1969 and 1970) that there are statistical similarities between a central place system and the rank-size distribution.

[8]It should be noted, however, that there is possible conflict between von Boventer's findings and the earlier empirical study of spatial aspects of urban growth in the United States by Madden (1958-9). Madden found that the growth rates of urban centres declined with distance from the metropolis up to 45 miles, grew steadily in the 45-64 miles range (also urban centres in this range tended to be larger than those nearer to or farther from the metropolis), declined again up to 114 miles, and were high but unstable beyond 115 miles. The implication of this is that there could, in fact, be an optimum distance where hinterland effects are strong but the competitive pull of the large centre very weak. This raises two questions. Does the distance over which a major metropolis exerts agglomeration effects differ from the radius of its competitive pull? Would hinterland effects be reduced by a location further away from the metropolis than shown in Madden's study because of falling population densities? Answers to these questions might reconcile the potential conflict.

[9]It is true that attempts to deal with the problems of metropolitan concentration, size and interurban distance were made in both The Strategy for the South East (1967) and Strategic Plan for the South East (1970). However, these are regional rather than national strategies, and in any event have not yet been implemented.

[10] This does not mean that the equilibrium metropolis policy is very successful; merely, that it implies a more clearly discernible national urban strategy.

[11] A variant of this might be the promotion of income equity within cities. I have not discussed this for two reasons. First, I would argue that the efficiency of the city as an instrument of national economic growth depends on inequities. Second, it is arguable that appropriate policies for dealing with extreme income inequalities are primarily national welfare and social measures rather than part of an urban policy.

[12] For example, in the United States an important social goal would refer to attaining desirable class-status-race mixes within cities of different size, and this would encompass still other goals, e.g. relating to income redistribution, education, labour markets, etc.

[13] Dissenters from this view may argue that reducing the size of the metropolis will not only raise the welfare of those induced to locate elsewhere but also the welfare of all remaining households and firms in the metropolis. This may be so, but I doubt whether social costs are easily reversible.

1 June 1971

The Agglomeration Process in Urban Growth

by

Joel Bergsman
Peter Greenston
Robert Healy*

NOTE: This paper has been prepared for the Resources for the Future - University of Glasgow Conference on Economic Research Relevant to National Urban Development Strategies, Glasgow, Scotland, August 30 - September 3, 1971. It is subject to revision and should not be cited or quoted without express permission of the authors.

* The authors are with The Urban Institute, 2100 M Street, N. W., Washington, D. C., USA.

JOEL BERGSMAN

Joel Bergsman has been on the Senior Research Staff of The Urban Institute since 1969 doing research on urban economic development. Before that he was Associate Research Economist and Lecturer in Economics at at the University of California at Berkeley. He was also with the University of California Brazil Development Assistance Program in Rio de Janeiro in 1966-1967 as Visiting Associate Research Economist and the Office of Program Coordination of AID in 1963-1965.

Bergsman received his Bacherlor's Degree in 1959 from Cornell in Electrical Engineering, and his Master's and Doctorate's Degrees from Stanford in business economics.

Articles and books by Bergsman include "Electric Power Systems Planning Using Linear Programming," published in Transactions on Military Electronics, No. 2, April 1965; "An Almost Consistent Intertemporal Model for India's Fourth and Fifth Plans," *Economic Weekly*, Bombay, November 1965, which was reprinted in *The Theory and Design of Economic Development*, Irma Adelman and Erik Thorbecke (eds.), Johns Hopkins Press, 1966; *Brazil's Industrialization and Trade Policies*, Oxford University Press, 1970; and "Alternatives to the Non-Gilded Ghetto," *Public Policy*, Spring, 1971.

PETER GREENSTON

Peter Greenston joined the Research Staff of the Urban Institute last year; prior to that he was with Robert R. Nathan Associates in Washington. He has a bachelor's from Oberlin College (1964) in Economics, and is presently a candidate for a doctorate from the University of Minnesota in Economics, expected in the fall of 1971.

ROBERT HEALY

Robert Healy is an economist with the Urban Institute, Washington, D. C. He is presently a Ph.D. candidate in economics at the University of California, Los Angeles. He has previously done research on the place of housing investment in economic development with the International Housing Productivity Study, UCLA. Among his research interests are the economics of growth centers and new cities and the role of agglomeration economies in industrial clustering and urban growth.

Bergsman
Greenston
Healy

This is a preliminary draft of a paper to be presented at the Resources for the Future-University of Glasgow Conference on Economic Research Relevant to National Urban Development Strategies, Glasgow, Scotland, August 30 - September 3, 1971. At the time of writing we are just getting the first preliminary and partial results in our research. The present draft therefore describes our objectives and methodology, and presents some of these results for illustrative purposes only. A revised version including more complete and reliable results will be distributed at the Conference.

The research is supported by funds from the U.S. Department of Housing and Urban Development. The views in this paper are those of the authors and do not necessarily reflect those of The Urban Institute or its sponsors.

1 June 1971

THE AGGLOMERATION PROCESS IN URBAN GROWTH

by

Joel Bergsman
Peter Greenston
Robert Healy

I. Agglomerating Forces

Introduction

It is commonly asserted that new technologies have made firms increasingly 'footloose', so that location decisions are made for amenity or other 'non-economic' reasons. Although this may be true for some standardized manufacturing processes, most economic activities locate not in isolation but near other activities in large urban agglomerations. Despite pollution, congestion, high taxes and high land costs, the metropolitan area has remained the preferred site for the fastest growing economic activities, particularly in the service industries. Few other cities offer the cultural attractions, the financial markets, the airline connections, or the variety of consultants and laboratories which are found in New York, or on a lesser scale, in the half dozen largest American cities. Even though many specialized services are only a small proportion of a firm's total purchases and hence do not loom large in input-output coefficients, they may be essential to a firm's survival. As such traditional locational factors as transportation and power have become more ubiquitous, the metropolis has retained its attraction, capitalizing on its role as a rich source of information and professional talent.

1 June 1971

Bergsman
Greenston
Healy

The research reported here views urban growth and decline as a constant agglomeration and deglomeration of economic activities, responding to external economies and diseconomies created by previous location decisions by firms and individuals. Initially, we are studying the effects of these forces in cross-section by studying the economic structures of cities at one point in time; later we will apply the insights obtained to studying spatial shifts of activities, analyzing how the structures shift over a period of years.

Agglomerative forces, although fundamental to the urban growth process, are not well understood. A really comprehensive theoretical statement is yet to be achieved and empirical studies are far from satisfactory. Our own research is heavily empirical. It seeks to synthesize existing theory by combining the insights of location theory and central place theory, and to test the resulting eclectic propositions with better data than has heretofore been used. Particular attention will be paid to new methods of identifying industrial complexes, to the relation of economic structure to the growth rate and to other characteristics of cities, and to the role of non-manufacturing industries in producing agglomerations.

In the U. S. today there is growing dissatisfaction with this concentration of people and activities, and growing discussion of the desirability of more 'balanced' regional distribution. Policies which attempt to change these patterns, however, must be based on a thorough understanding of the forces which have produced them. Our research is meant to contribute to this understanding. This paper outlines

1 June 1971

the research strategy we are following, shows how it will be a useful input to formulating urban growth policy and presents some illustrative early empirical results.

Concepts of Agglomeration

The literature divides agglomeration economies into 'localization' and 'urbanization' economies. Localization economies result from the proximity of several firms engaged in the same activity. They help explain such specialized concentrations as jewelry in Providence, style apparel in New York, and aircraft in Los Angeles. The availability of skilled labor and specialized business services, the ready exchange of information and concentrations of customers make sites where production in an industry is already taking place particularly attractive as the location of a new firm. Moreover, there is some tendency for innovations in a given industry to occur near the site of current production, giving rise, for example, to the small firms 'spun off' by existing firms in the California electronics industry.

Urbanization economies create two other types of clustering. The first is the clustering of firms in different but related industries, locating together in industrial complexes such as oil refining/petrochemicals, food processing/container manufacturing, and metal fabricating/machinery. The linkages which hold these firms together are input-output flows between firms. At times these flows are direct, with one firm serving as the supplier or customer of another. In other cases, two firms will be linked together spatially because of their common relationship with a third.

The second type of urbanization economies causes both individual firms and the industrial complexes described above to locate in or near large cities. For example, over ninety percent of national employment in such industries as lithographing, paints and varnishes, periodicals, electrical control apparatus, millinery, and motor vehicle manufacturing is located within metropolitan areas. Style apparel and advertising are both concentrated in New York. But probably they do not attract each other; rather each is there because of some other characteristics of the city, such as its primacy in communications media and its role as a style setter for the nation as a whole. Clearly, the attributes of metropolitan areas are more important for some types of activities than for others. Along with the external economies, the metropolis presents many disadvantages, many of them a direct result of the clustering process itself. Thus we observe the tendency of the textile industry to follow two location patterns with standardized manufacturing of goods such as bedding and underwear moving to low wage, non-metropolitan locations in the South and Middle West while designing of high quality fabrics or high fashion goods remains within the metropolis.

Alternative Research Strategies

Agglomeration economies were first discussed by Alfred Weber in his classic book on industrial location theory. Subsequent work by Ohlin, Robinson, Hoover and others provided a typology, distinguished economies due to technical complementarity from those associated with city size, and pointed out that agglomeration economies are a spatially limited form of Marshallian external economy. (Weber, 1929; Ohlin, 1933; Robinson, 1932; Hoover, 1937, 1949.) The few mathematical formulations have concentrated on models of two or three firms with relatively simple interrelationships. Isard (1956, 1969) added dynamic elements to Weber's 'critical isodapane' model by proposing game theoretic solutions in which firms bargained over the distribution of potential gains from agglomeration. Dunn (1970) and Meier (1970) have recently borrowed from information theory in seeking more general models of the interrelationships among firms in urban areas.

On a much more highly aggregated level discussions of the optimal city size problem have recognized that agglomeration economies cause per capita product to rise over some range of city size, offsetting the rise in per capita costs. (Klaassen, 1969; Alonso, 1970) Unfortunately, such studies offer few insights into how these advantages are produced, or how their effects vary among different types of firms.

Some of the most interesting work in the study of agglomeration consists of descriptions of clusters in particular cities. Most of the speculation and hypotheses which they offer for the existence of these clusters are as yet untested. The New York Metropolitan Region Study presented innumerable examples of how many of the area's industries depend on external economies. (Hoover and Vernon, 1962; Lichtenberg, 1960; Vernon, 1960) Chinitz (1961) claimed that the relative availability of externalities explained the differential growth of Pittsburgh and New York, emphasizing the role of innovation and of capital markets. Florence (1955) outlined the advantages of large cities in providing a wide range of services and cultural amenities, while Jacobs (1969) re-emphasized the role of the metropolis as an incubator of new industries.

There have been a few attempts to measure directly the cost savings due to agglomeration. Isard, et al. (1959) developed the technique of industrial complex analysis to choose among alternative packages of technically related activities. Shefer (n.d.) and Rocca (1970) found that the degree of overall industrialization in a region contributed significantly to higher output per worker in selected manufacturing industries. Further progress in direct measurement of externalities requires better data on capital per worker than are now available. Without better data, other methodological refinements or variations will probably not add much to our understanding.

As this discussion indicates, much theoretical work in this field has been microanalytic. Including explicit assumptions about the behavior of decision-making units has a natural appeal. However, there are two limitations of research based directly on such models. First, the models often are based on particular assumptions about the objective functions and production functions of the business firm that may not include some of the forces which influence location. The hypotheses generated by these models should be among those tested in explaining results of actual location decisions, but limiting testing to one or a few such models increases the risks of misinterpretation due to spurious correlation or loss of significance in the results.

The second problem for us stems from our wish to produce results which are useful in formulating urban growth policies. Few of the target variables in a national urban growth policy appear in microanalytic models. Some of these target variables can be thought of as aggregations of variables in the micro models. (E.g., the size and economic structure of a city are aggregations of the size of each firm in it.) But in practice the aggregation cannot be done 'after' the model; rather, the model itself must deal with the aggregate variables.

An alternative to microanalytic models is empirical analysis of the location patterns resulting from all the individual decisions. There is a large body of such research on the localization, clustering and urbanization of firms. We believe that most of this research can be improved upon. Geographers have pointed out patterns of localization, without analyzing the role of agglomeration in cost reduction (Nelson, 1955; Alexandersson, 1965; Murphy, 1966) Economists, who might be expected to be more attentive to external economies as a location factor, have tended to concentrate their work at the state or regional level, where data are more tractable but less useful. (Perloff, 1960; Fuchs, 1962) Richter (1969) and Streit (1969) have studied clustering in manufacturing, without considering the interaction of manufacturing establishments with those in the service and other supporting industries. Most recently, Stanback and Knight (1970) have described how, for highly aggregated sectors, the degree of clustering varies with a city's size and position in the functional hierarchy. They do not analyze the reasons why the various clusters exist as they do, and indeed probably could not with the high degree of aggregation of their data.

II. Research Design

The direction of our own approach to the agglomeration problem should be evident from our discussion of the shortcomings of the existing literature. We believe that we can improve on previous work in several important ways. While most empirical work has concentrated on manufacturing, urban growth increasingly depends on the service industries. We will pay particular attention to services and to government employment. Aggregation of diverse industries tends to obscure clustering behavior -- our degree of industrial detail will be far greater than in previous studies.

There are many hypotheses about the location of activities, many of which postulate some form of agglomeration or clustering process. Our first step is to define a set of activity clusters, and the second step is to test hypotheses about their structure, location and interaction among themselves and with other characteristics of cities.

Data

We will measure the structure of metropolitan economies by employment. An output measure would be preferable, but on the level of detail desired only employment data are available. We have assembled a file for 1963 consisting of employment in 203 SMSAs, with 3-digit SIC level detail in manufacturing and 2-digit detail in trade, agriculture, mining, construction, government and service industries (186 industries). Later work will deal with 3 and 4-digit disaggregation for 1965 and 1970, with the sample enlarged to include selected non-SMSA counties.

The choice of the unit of observation is crucial for our research, involving more than definitional concerns. The county is the basic statistical unit for much of the data we will use. Our choice in defining units is therefore limited to determining appropriate aggregations of counties.

The most well-known aggregation is the Standard Metropolitan Statistical Area (SMSA). Essentially, a map of SMSAs separates the country into 'metropolitan' and 'non-metropolitan' categories, with the former currently divided into approximately 247 areas. The criteria used to define SMSAs are arbitrary, however, in both the minimum size criterion and in the criterion for including adjacent counties. Our research is concerned with interactions in markets for labor, materials, and services, many of which may extend beyond the borders of an SMSA. Moreover, independent regional centers of less than 50,000 population are the hubs of labor markets in the less densely settled sections of the country, paralleling in their role centers of greater population where settlement is more dense.

Adding adjacent counties to SMSAs is possible for us, but will probably not be done unless obvious cases arise in which the SMSA as defined clearly excludes an adjacent county which is a significant part of the metropolitan area. However, the functional economic areas defined by the U. S. Office of Business Economics give us some good guidelines on combining adjacent SMSAs which are in fact parts of one economically integrated metropolitan area. Our procedure, then, will be (1) to use the SMSA as the basic unit of study and the county as the basic building block; (2) to combine two or more SMSAs when they appear to be integral to a greater complex; and (3) to include additional areas whose centers are too small to be SMSAs but in which industrial activity is not insubstantial. Fortunately, the data permit experimentation with the unit of observation, and we intend to do such experimentation. One technique which should suggest changes in our definitions of units is analysis of residuals;: if a particular city frequently appears far off regression lines, its definition might be improper.

An alternative approach which circumvents the area delimitation problem is to measure variables in terms of geographic potential. (Isard, 1960) The measure is essentially one in which the 'potential' of the phenomenon in question at a given point is depicted as a function of the phenomenon at that point and at all other points. The flexibility and generality of the potential approach make it attractive, and we intend to experiment with it.

Description of Industrial Clusters

The basic measure of association between industries is the correlation across metropolitan areas, of employment in each industry with employment in every other industry.* To correct for the correlation of the absolute

---

* There are several other measures of industrial association, the best known of which is P. Sargant Florence's "coefficient of geographic association." (1943) The superiority of the simple correlation coefficient has been pointed out by McCarty (1965) and Richter (1969).

level of employment with city size, employment in each sector is converted to a per capita or share-of-total basis. This simple correlation analysis results in an n x n matrix of correlation coefficients between the n industries considered. We realize that even when industries are strongly interdependent employment in one will not necessarily vary linearly with employment in another. Stanback and Knight (1970), for example, tested whether cities with employment in the highest quartile in one of a pair of industries were also in the highest quartile for the other. We plan to experiment with several measures of this kind, including rank order correlation coefficients. We are also making use of the Automatic Interaction Detector program, which identifies association with minimal prior specification of functional form. (Sonquist and Morgan, 1964)

Industries locate in clusters containing not just two, but varying numbers of economic activities. Factor analysis is ideally suited to identifying such groups or clusters. We use factor analysis to identify groups of industries which show similar location patterns and then test hypotheses about the causes of these groupings with more conventional multiple regression techniques.

Types of Clusters and Mechanisms of Linkage

One obvious hypothesis is that the spatial clustering of firms follows market linkages; that is, firms locate near their suppliers and their customers. This hypothesis will be tested with the aid of the 478-sector input-output table for 1963, with some sectors aggregated to match the employment data previously described. From this table we may derive an n x n matrix of 'transaction coefficients' which describe transactions between pairs of industries as a percentage of their total sales. One way to calculate a transaction coefficient is:

$$t_{ij} = \frac{s_{ij} + s_{ji}}{S_i + S_j} \qquad (1)$$

where:

s_{ij} = sales of industry i to industry j

S_i = total sales of industry i

We will then compare the two matrices, one containing a measure of spatial association (r_{ij}, the simple correlation of employment in pairs of industries) and the other a measure of technical association (the transaction coefficient) for each pair of industries. The industrial pairs may be separated into five or six groups, corresponding to the cells of the following contingency table:

<table>
<tr><th></th><th>Spatially Linked, Positively</th><th>Not Spatially Linked</th><th>Spatially Linked, Negatively</th></tr>
<tr><td>Technically Linked</td><td>Agglomerating Industries (r_{ij} high and positive, t_{ij} high)</td><td>Non-Associating Industries (r_{ij} low, t_{ij} high)</td><td rowspan="2">"Repelling" Industries (r_{ij} large and negative)</td></tr>
<tr><td>Not Technically Linked</td><td>Associating Industries (r_{ij} high and positive, t_{ij} low)</td><td>Non-Linked Industries (r_{ij} low, t_{ij} low)</td></tr>
</table>

In addition to technical linkages through buying from and selling to each other, firms may also be linked by common needs for certain labor skills. We will attempt to collect detailed data on this, and to investigate whether these commonalities are associated with geographical concentration for various industries.

Agglomerating industries are those pairs which have strong input-output links and which juxtapose spatially. The oil refining industry and the petrochemical industry are probably in this class, as are central office employment and data processing. Associating industries are found in similar locations, but have no direct links. We might distinguish two classes of these. One consists of industries which have common locational needs. For example, the motion picture industry and the airframe industry are both attracted to areas of mild climate, but have virtually no input-output linkage. A second group of associating industries is made up of those industries which have indirect input-output links. These industries may share common suppliers or common customers, yet have no transactions with each other. Industries of this type will be discussed in more detail below.

Non-associating industries are those which, despite strong economic links with other industries, fail to cluster spatially with them. Generally, this is due to a situation in which each industry of the pair is subject to a different, and overriding, locational force. An example would be grain milling and bakery products. Grain milling is oriented to the location of grain production and to rail and water transport. Bakery products, on the other hand, show a strong final market orientation.

Industries classified as 'non-linked' are probably those in which unskilled labor input is dominant and which seek non-metropolitan locations with abundant low-wage labor. Such an industry would tend to avoid existing industrial concentrations with their relatively high wages.

A few groups of industries may display negative attractive forces -- they may 'repel' each other. Inclusion of additional explanatory variables later in the analysis may reveal more of such interactions.

Economic links between industries are more subtle than the volume of direct transactions. Two industries, each of which sells to a third industry, are also linked. They may form a three sector industrial complex, as shown in Figure 1 below (arrows indicate the direction of commodity flows):

Figure 1 Figure 2

The ABC complex in Figure 1 will cause the same regional employment mix as that of Figure 2, despite the fact that the linkages are quite different. The classification described earlier will account for this difference in part, for industries B and C of Figure 1 would be classified as 'associating' and B and C of Figure 2 as 'agglomerating'.

<u>Type of City Which Attracts Each Cluster</u>

Thus far, we have outlined an investigation of the observed tendency of groups of firms to locate in the same metropolitan area. Another phenomenon we intend to study is the variation of the composition of economic activities among different cities. What kinds of clusters occur in what kinds of cities? Central place theory has given us some tantalizing insights into the mechanism of the urban hierarchy, but has neglected interrelationships of firms within the central place.

We will attempt to identify the size and other characteristics of cities which have high concentrations of particular industries. The clusters of spatially related industries derived from the factor analysis described earlier will be useful here. Each city will have a factor score indicating its concentration in those industries composing each spatial cluster. By investigating the relationship of factor scores to the size and other characteristics of cities, we can identify the propensity of different clusters to locate in cities of different types.

There are a variety of ways in which city size affects the composition of activities. Large cities provide a market large enough to permit a great deal of specialization -- activities which in a smaller city would be performed within the firm are contracted to independent organizations. Moreover, the large market makes it possible to provide services which are not produced at all in smaller centers. Many observers have suggested that large cities provide a climate conducive to the development of new industries. We would therefore expect to find that those industries which are concentrated in large cities are, on average, 'newer' than those which are found in smaller centers.

The effect of city size on the industrial structure may depend on the proximity of the city to larger cities. Especially for service industries where scale economies are important, central place theory would lead us to expect that an SMSA of, say, 300,000 people in northeastern Pennsylvania would offer very different advantages than one of the same size in North Dakota. Even for manufacturing such effects are probably significant due to transport costs. We will test several measures of position within the urban system, such as distance to nearest larger city, distance to nearest city over a given size, population potential and economic potential. The relationships may not be linear, and various forms should be tested.

Another variable, or group of variables, relates to the city's labor force. Among cities, labor differs in cost, in productivity, in education and training, in availability and in racial and sex composition. Industries differ in their labor requirements along some of the same dimensions. Low wage industries such as textiles, apparel and electronics assembly have changed their locations as labor costs changed.

Other obvious location variables have to do with amenities. (Ullman, 1954; Perloff et al., 1960; Berry and Neils, 1969) Just a glance at a map showing differential growth in different parts of the country during the last ten or fifteen years will show that indices of seasonal variations in temperature, annual rainfall, etc. will explain a large part of regional differences in growth. Moreover, the idea of the growth of 'footloose' industries, in many of which success for a firm depends on being able to attract the cream of highly skilled types of personnel, provides a rationale for considering the amenity factor to be a real cause, rather than a proxy for something else. Including an amenity index as one explanatory variable should permit us not only to measure its influence, but to determine more accurately the influence of other factors. For example, low wage rates are probably somewhat correlated with climatalogical amenities -- roughly on a north-south basis -- but we expect that enough difference exists -- roughly east-west -- to distinguish the two effects. The effects of amenities will obviously vary among industries, and our relatively high degree of disaggregation should provide new information about this variation.

We may encounter a number of metropolitan areas which do not fit the general patterns. Washington, D. C. is probably such a area, and some state capitals may be also. The Greater New York area may be a different kind of special case. Systematic analysis of residuals will be necessary to suggest such cases, which will then require either additional explanatory variables or separate treatment.

Multiple regression analysis will be used to learn about the type of city in which each cluster occurs, and about any ways in which the precise nature of the cluster depends on characteristics of the city. Such analysis can also be used to try to separate the effects of size as such from the effects of the industrial mix.

The type of city in which a cluster occurs can be studied by a single equation multiple regression:

$$c_{ij} = f(P_i, Z_i) \quad (2)$$

where c_{ij} = employment per capita in cluster j, in city i; alternatively, the factor score of city i on cluster j

P_i = population or other measure of economic size of city i

Z_i = other characteristics of city i, including availability of transport, accessibility to other cities, amenity index, labor force composition, etc.

The size of the city might be measured by population, or alternatively by value added produced within it.

In the simplest model, employment in each separate industry within the cluster can also be thought of as determined by the characteristics of the city:

$$n_{ij} = f(P_i, Z_i) \qquad (3)$$

where n_{ij} = employment per capita in industry j, in city i.

However, we intend to try to identify separately the effects of the other industries in the cluster, characteristics of the city, and perhaps industries not usually associated with the cluster:

$$n_{ij} = f(c_{ij}-n_{ij}, \bar{n}_{ij}, P_i, Z_i) \qquad (4)$$

where n_{ij} = employment per capita in industry j (a part of cluster c_j), in city i

$\bar{n}_{ij}$ = employment per capita in other industries j, <u>not</u> parts of cluster c_j, in city i

A positive coefficient on $\bar{n}_{ij}$ in equation (4) would mean that certain activities $\bar{n}_{ij}$ may not usually occur with n_{ij}, but that in particular types of cities (as represented by P_i and Z_i), the two activities do tend to associate. Analogously, a negative sign on P_i or Z_i would mean that the particular city characteristic P_i or Z_i 'substitutes' for n_{ij} in the cluster.

Equation (4) can be expected to present a number of econometric problems. First, in some instances we are finding that $c_{ij}-n_{ij}$ is too highly correlated with n_{ij} to allow detection of any other effects. In these cases we must use equation (2) for the cluster as a whole. Also, both $c_{ij}-n_{ij}$ and $\bar{n}_{ij}$ in (4) are endogenous variables, possibly influenced by n_{ij}, P_i, and Z_i. Whether simultaneous equation methods can be used will depend on the precise formulations we adopt, which Z_i we think are important in each equation, etc.

Still another variation in the above analysis would be the use of absolute employment as a dependent variable. For many industries, scale may not vary with city size, and n_{ij} will be large in small cities and small in big cities. For such industries, the absolute amount of employment may be a more appropriate measure of the presence of the industry.

The Range of Agglomeration Economies

Even less is known about the range of agglomeration economies than about their composition. Some may be obtainable only by contiguous location, such as the location of a steel furnace which receives molten iron from a blast furnace. Others are limited by the size of labor markets, and may be obtainable at sites within, say, two hours driving distance from each other. The concentration of manufacturing in the so-called Old Manufacturing Belt suggests that some agglomeration economies may be effective for distances of several hundred miles. By investigating the variation in degree of clustering as the level of aggregation is varied, the range of the agglomerative forces may be tentatively identified.

We do not know, *a priori*, the geographic range over which agglomerative forces operate. Choosing the metropolitan area as the unit in which industrial clusters are defined has much appeal -- counties are clearly too small, and states are clearly too big. Any choice, however, amounts to assuming the answer -- choosing the metropolitan area means we assume that agglomerative forces do not vary with distance within a metropolitan area, but drop to zero across metropolitan boundaries.

For example, Pittsburgh ranks high in both primary steel and fabricated structural metal products. Akron, high on the latter, produces almost no primary steel, yet because of its accessibility to Pittsburgh, Youngstown, Buffalo and other important primary steel centers, has a high 'potential' relative to steel producers.

Potential measures accessibility of a city to a given activity throughout the country by adding up the amount of a given activity in each other city in the country. The amount for each city is divided by some function of the distance between that city and the city for which the potential is being calculated. In algebraic terms, continuing the example,

$$S_{Akron,\ steel} = \sum_i \frac{N_{i,\ steel}}{f(d_{i,\ Akron})} \qquad (5)$$

where $S_{Akron,\ steel}$ = potential of Akron to steel manufacturing

$N_{i,\ steel}$ = employment in steel, city i. (Alternatively, N can be any variable such as population or disposable personal income.)

$f(d_{i,\ Akron})$ = a function of the distance between Akron and city i,

and the sum is over all cities (metropolitan areas) i.

This formulation avoids much of the arbitrariness of any fixed boundaries. The distance decay function f(d) can be estimated from the data (sometimes directly, or at least by trial and error) rather than arbitrarily assuming that f(d) = 1 for d = 0 and f(d) = infinity for d>0, as is implicit in the non-potential formulation.

Changes in Size and Structure of Urban Economies

The way an urban economy grows or declines depends on its present size and structure. For example, there has recently been a tendency for medium sized cities to grow faster than those which are very large or very small. Moreover, changes in the industrial composition of national output have different implications for specialized manufacturing cities than for "regional capitals".

We choose to approach this problem through the framework of a 'shift-share' analysis. (Perloff, et. al., 1960; Borts and Stein, 1964; Berry and Neils, 1969) The growth of employment in a particular industry in a particular city can be broken down into three components: overall national growth in employment (r); the difference between national growth in the particular industry (r_j) and overall national growth, and the difference between local growth in that industry (r_{ij}) and national growth in that industry:

$$\frac{\Delta N_{ij}}{N_{ij}} = r_{ij} = r+(r_j-r)+(r_{ij}-r_j) \tag{6}$$

where N_{ij} = employment in city i, industry j, in the initial year. Omission of a subscript indicates the sum over the omitted subscript.

r_{ij} = percentage growth in employment during the period in city i, industry j.

Δ = indicates change during the period

$n_{ij} = \frac{N_{ij}}{N_i}$

A little algebra gives an analogous equation for the growth of employment in the city as a whole:

$$\frac{\Delta N_i}{N_i} = r_i = r + \sum_j n_{ij} \ [(r_j-r)+(r_{ij}-r_j)] \tag{7}$$

Equation (7) says that the growth of a city's economy can be expressed as a combination of (exogenous) national factors (r and r_j), the city's (given) initial economic structure and size (n_{ij} and N_i) and the difference between the local growth rates and the national growth rates for each industry in the city ($r_{ij}-r_j$). Only the last of these three terms is endogenous to a model of changes in the city's economy. It is the forces which affect this variable -- which has been called the competitive shift -- that we wish to study. Why does a particular industry grow faster in some cities, and slower in others?

The work on agglomeration, described above, will help in pinpointing hypotheses to test in explaining changes. Where that work analyzed the economic structures of cities at a given time, this will analyze changes over time.

The most obvious variable to examine in explaining the competitive shift is the size of the city. Theories and observations, from Adam Smith through early location theory to Jane Jacobs, lead us to expect that certain activities will do best in cities of certain sizes.

The oldest thought on this centers on economies due to the presence of other firms. This has already been discussed. Our work on agglomeration and industrial clusters should help us to separate the influence of structure from that of city size as such. We can use the size of other industries which are shown to be associated geographically (and/or through input-output or analogous linkages) as explanatory variables for the competitive shift in a given industry. For a number of industries, especially in the manufacturing sector, we may be able to show that city size as such does not affect the competitive shift, if the presence or absence of certain other industries is separately accounted for. For other industries, notably services and some manufacturing where relationships with particular customers or suppliers are less important, but internal scale economies are important, city size as such may be the only way to represent economies of urban concentration.

On the other hand, there are some reasons to expect competitive shifts to be inversely related to size of the industry itself. Borts and Stein (1964) report that growth in manufacturing was inversely related to the share of manufacturing in total output (using states as units of observation, during 1919-1953). Thompson (1965) notes that the fifteen largest metropolitan areas became more alike in their industrial structure (within a roughly one-digit breakdown) between 1950 and 1960. That is, industries grew more rapidly where they were less highly concentrated. This phenomenon is sometimes explained by a theory of diffusion. In the early stages of development an industry tends to cluster, both because it started in only a few places and also because it needs much 'information' and specialized inputs which are not available elsewhere. When the industry matures, the argument goes, its operation becomes more routinized and it can disperse toward other markets, low-cost inputs, or other advantages which might have been less important for its inception. (Thompson, 1968; Jacobs, 1969)

Our research design should permit improved testing of these ideas. For testing the diffusion theory, the age of an industry or its phase of development is measurable; if not as a continuous variable, as a set of dummy variables. The size of the city will be included separately in the regression, and so can a measure of the age of the city.

In the same way, we can determine the influence on growth of such factors as transportation, the availability of labor, location within the urban hierarchy (which is not always coincident with city size), amenities, and others.

Because of the obvious data limitations, we will not be doing time series analysis. We intend to analyze changes in city size and economic structure over a period of time, again using cross-section data. The first comparison will involve changes between 1965 and 1970.

III. Policy Options and The Research

Intellectual curiosity is not our only motivation in this research. Many Americans are concerned with patterns of urban growth, or more accurately, with problems which those patterns seem to produce or exacerbate. The Housing and Urban Development Act of 1970 finds that:

> "...the rapid growth of urban population and uneven expansion of urban development in the United States, together with a decline in farm population, slower growth in rural areas, and migration to the cities, has created an imbalance between the Nation's needs and resources ...
>
> "...the Federal Government ...must assume responsibility for the development of a national urban growth policy which shall incorporate social, economic, and other appropriate factors. Such policy shall serve as a guide in making specific decisions at the national level which affect the pattern of urban growth..."

The pattern of urban growth in the United States is a result of a hodgepodge of federal, state, and local programs and private decisions, whose ends and means are frequently in conflict and which often exacerbate some problems while ameliorating others. Growing concern for the quality of life, and the particular problems of the disadvantaged, have increased public belief that patterns of growth should be changed.

The principal sources of dissatisfaction can be broadly divided into two categories. One is largely intra-metropolitan in scope, and relates to fiscal disparities between central cities and suburbs, racial and economic segregation, urban sprawl, congestion, etc. The other category is inter-metropolitan, and relates to the growth of gigantic 'megalopolises,' which some see as increasing administrative unwieldiness, pollution, congestion, social discontent, and environmental ugliness, and the corollary fact that many small towns and rural areas are losing population and becoming ever less attractive places to live and work.

Our research is addressed to the second of these concerns.

A number of different strategies to deal with these trends have been proposed. Four examples show their range:

1. A large-scale new cities program, such as the 100 new cities averaging 100,000 population, and 10 of at least one million population, proposed by the National Committee on Urban Growth Policy and recently endorsed by David Rockefeller.

2. A growth center program aimed at small depressed towns and rural areas, such as past programs of the Economic Development Administration (EDA), preferred by many of the representatives of such places and by many others on grounds of equity and/or respect for small-town and rural values.

3. A growth center program aimed at increasing the growth of already prosperous, medium-sized cities, preferred to the two previously mentioned strategies by many economists and geographers on grounds of efficiency and indeed of feasibility.

4. A less place-oriented set of programs, such as some mix of income maintenance, education and manpower training, and revenue sharing with existing state and local governments on a population-income basis. Such a strategy differs from the first three in that it accepts whatever spatial concentrations emerge and concentrates on assistance to people and/or governments in these places.

Actual strategy could be any mix of these four alternatives, subject to important budgetary limits which force the strategies to compete for funds as well as for the attention of policy-makers. In the U. S. today the actual strategy is virtually a pure choice of the fourth. The government has made no explicit choice relating to a given distribution of city sizes and/or population, and there are few if any significant programs designed specifically to affect such variables. (EDA, which has tried to affect outmigration from the Appalachian region without much success, now has less funds than previously and faces an explicit proposal from the President for its extinction.)

The Nixon Administration has proposed a 'special revenue sharing' program which would allocate about $2 billion (in the first year) to cities, 80 percent of which would be allocated to cities in the 247 SMSAs according to formulas designed to measure need. Part of the other 20 percent would go to non-SMSA 'growth centers.' An additional $1.1 billion for 'rural areas' (this includes cities up to 50,000 population) would also be allocated by formula and an extra $100 million is provided expressly for cities in the 20,000-50,000 range. The rhetoric surrounding the proposals makes much of a growth center strategy, but it is an open secret in Washington that the principal criterion used in designing the formulas was "How do we re-elect Republicans in 1972?" Moreover, the proposals are interesting mainly as expressions of the Administration's intent. They are not likely to be passed by the Congress.

It is not clear that lack of a place-oriented strategy is wrong. First, under any conceivable priorities for new towns, growth centers, or rural development, far more resources would still be required for the needs of people where they now are and will remain. New towns may be nice experiments, and the several strategies might reduce the growth of many existing cities to some extent. But most Americans will still live in our existing cities, and problems of congestion, pollution, inadequate housing and schooling, poverty, and racial discrimination will not be eliminated by building some new towns, promoting some growth centers, and developing some rural areas.

Second, there is not strong evidence that the current geographic distribution of people and jobs departs greatly from the optimal. There is a lot of talk about external diseconomies, but many of the "externalities" may be expressed in market prices, and others may partially or even largely be balanced by external economies.

Our knowledge of the determinants of urban growth and decline is insufficient to foresee the results of present policies, much less to design programs which would give any assurance of net improvement. Our research is set in the context of these concerns, policies, and lack of policies. It is designed to learn more about how and why cities grow or decline. This will permit better estimation of the likely effects of any proposed programs to affect that behavior. Our research will say but little about the desirability of such effects; we aim mostly to learn more about the behavioral characteristics of the system to be acted upon. We will not analyze explicit policy alternatives; rather, we will try to define the 'feasible space' and the limits beyond which policy cannot expect to succeed. Our goals are precise and detailed descriptions of groups of economic activities which cluster together, of the economic structure of metropolitan areas, and of the specific spatial, functional, and geographic characteristics which are associated with a particular economic structure. This should enable us to distinguish between consistent and inconsistent goals for the growth of a metropolitan area in terms of size, economic structure, and place-fixed characteristics.

A few examples will show more clearly how our results might be useful. "Free-standing new cities" present real problems of economic viability. These new cities are meant to have more-or-less balanced industrial-residential mixes and not to be suburbs or satellite cities. But what kinds of industries should or will locate in one, and what measures would be needed to attract desired industries and not undesired ones? For a given targeted size and a given place, our results could be used directly to predict the 'expected' industrial mix. This would be a complete description of the number of employees in each different type of manufacturing, services, trade, government, etc., on a 3-to-4-digit level of detail. Some further research could then rather easily go from these results to expected socio-economic characteristics of the population. The planners of the city could compare these results with their objectives, revise their plans if and as appropriate and then take steps to attract activities which they feel are too little represented in the expected mix, and to repel other activities that are not wanted but would otherwise appear.

Similar analysis could be done for growth centers -- small or medium-sized cities in which more rapid growth is to be induced. But perhaps even more important for a growth center strategy, our results could be useful in the earlier stage of selecting the centers. Most simply our results could directly predict the growth rate for any metropolitan area. Those areas which are not expected to grow rapidly would presumably be poor candidates for growth centers. More sophistication and flexibility could be built in to the analysis by searching for cities whose industrial mix has a large missing element of high-growth or otherwise desirable industries -- i.e., cities in which certain desirable industries are less important than our results predict. There would be a presumption that such 'missing' industries could be easily induced to move to those cities. Projected rural or small-town development programs could be evaluated and modified through analogous uses of our results.

Another potential application is to projection models. To the extent that the research fulfills our expectations and adds to understanding about why cities grow or don't grow, growth projections for particular metropolitan areas can be made more accurate. Conditional projections, showing the effects of exogenous or policy variables on the projected growth, could be made much richer because of the light our results should shed on what those variables are and the magnitudes of their influences. Our analysis of the effects of specialized services on city growth, and our hoped-for synthesis of the effects of the industry mix, locational factors affecting the competitive shift, and central place hierarchical factors should be especially valuable in improving projection models.

IV. Preliminary Results

The research design which we have described requires an enormous amount of data on each of the more than 200 SMSAs. We have now assembled a data file on the industrial structure in 1963, using employment figures from the Census of Manufactures, Census of Business, County Business Patterns, Census of Governments and other sources. This file covers 144 manufacturing industries (3-digit SIC detail) and 42 non-manufacturing industries (mostly 2-digit SIC detail). Geographically, the data cover all of the 203 SMSAs defined in 1963 in the mainland U.S.*

One of the most persistent data problems encountered when working at the SMSA level is the Census Bureau's disclosure rule, which prevents publication of employment and other data for any industry-place combination where the number of firms involved is small. In many industries, particularly in manufacturing, a small number of firms may account for thousands of jobs. Fortunately, the Census Bureau releases figures on number of plants by county and industry for each of 7 size classes. We have combined information about the number of plants by employee size class with data for mean number of employees per plant (for each industry nationally) for the same size classes and estimated the data which the Census Bureau

* For New York and Chicago, the Standard Consolidated Areas are used. These are aggregations of the central SMSA with one or more contiguous SMSAs. For several New England cities, SMSAs are defined on a township rather than a county basis. In these cases, we have used an alternative definition (Metropolitan State Economic Area) which is based on counties.

did not release. By adjusting our figures to meet regional employment totals, we are confident that our margin of error is small. We plan to improve even more on the estimation procedure as we construct the more detailed employment data files for 1965 and 1970.

Before presenting a selection of our empirical results, we must re-emphasize that they are a product of a very early stage of the analysis and represent more than anything a test of our research strategy and computer programs. They involve assumptions about the measure of industrial association, functional forms, and factor rotation which remain to tested further as the analysis proceeds. We present them less as a report of where we have been than as an example of where we are going. A revised and extended report of our results will be circulated at the time of the Conference.

We first followed previous studies in measuring the pairwise clustering of firms by the simple correlation of their (per capita) employment. Our 186-industry matrix contained 17,205 possible pairs. Even if there was no underlying correspondence in the location patterns of any of the industries, we would expect that seventeen correlations would be greater than .231 or less than -.231 (.001 level of significance). In fact, there were 1003 such pairs. Thus there seem to be far more geographical associations than would appear if industrial locations were determined by chance.

Many of the pairs reflect familiar patterns of industrial association. A common pattern is the association of an industry producing a material which is expensive to transport with another industry using that material as an input. Examples are petroleum refining/chemicals (r = .486); logging/sawmills (r = .921); and iron foundries/engines and turbines (r = .464). In other cases, employment in two industries is associated because both use low wage labor (leather footwear/woolen fabrics; millinery/ electronic components) or, alternatively, employ workers with high levels of skill (mechanical measuring instruments/photographic equipment). At times, location patterns of two industries coincide because both seek to locate in the same size or kind of city, as is the case for non-profit organizations and commodity brokers (r = .500) or for banking and advertising (r = .539).

Surprisingly enough, only thirty out of the 1003 significant correlations were negative. We had expected that there would be several types of industries which would 'repel' one another, that is, cities which rank relatively high on one of a pair of industries would rank relatively low on the other. Among them, we had Hypothesized, would be high wage/low wage pairs (motor vehicles/textiles); industries which locate in large metropolitan areas and those which locate in smaller ones (advertising/lumber) and the service industries and heavy manufacturing (retail trade/engines and turbines). Among the few industries which did show negative clustering behavior, only the last of these patterns was apparent. Excluding the negative correlations of "unclassified industries" with other sectors, the significant negative correlations were:

Industry Pair	r_{ij}
Federal government/medical services	-.255
Federal government/fabricated structural	-.237
Federal government/metal stampings	-.232
Medical services/misc. textile products	-.272
Medical services/weaving mills, cotton	-.264
Automobile repair/blast furnace and basic steel products	-.233
Personal services/paper mills	-.238
Personal services/blast furnace and basic steel products	-.255
Credit agencies/nonferrous rolling and drawing	-.234
Credit agencies/nonferrous foundries	-.233
Construction/metal stampings	-.282
Construction/nonferrous foundries	-.253
Construction/toys and sporting goods	-.253
Construction/motor vehicles	-.234
Construction/general industrial machinery	-.232
Construction/blast furnace and basic steel products	-.232
Retail trade/blast furnace and basic steel products	-.272
Insurance agents/blast furnace and basic steel products	-.254
Electric lighting and wiring eq./concrete	-.235
Concrete/misc. apparel	-.254

The use of pairwise correlations to describe agglomerations of activities imposes an arbitrarily simple and rigid structure on some very complex interindustry ties. Industries locate near one another not merely because of economies in transporting goods between stages of a production process. Firms in a dozen or more industries may have common location patterns which can be interpreted only when we can view them as a group or cluster. Moreover, a matrix of over 17,000 individual correlation coefficients is hardly convenient for hypothesis testing. A technique was needed which would more economically define industrial clusters containing several industries and which could separate those cities where a particular cluster was important from those where it was not.

Factor analysis was used to arrange the 186 industries into groups such that spatial association was maximized among the industries within each group (or factor) but minimized between groups. Common factor analysis was applied to the matrix of correlation coefficients. The resulting principal factor loading matrix was rotated to a varimax solution of fifty-seven factors. It was decided to rotate fifty-seven factors because a discontinuity between the 57th and 58th factors in their contributions to explained variance was observed. The rotated factor solution describes 57 clusters of industries, each of which is composed of industries which locate in the same places. The factor analysis reproduced 81% of the common variation in per capita employment in the 186 industries over the 203 SMSAs.

The factor which explained the greatest proportion of the common variance (about 5 percent) was composed almost entirely of service industries; particularly of the kind of services offered in the largest market centers. The single industry most closely associated with this factor was advertising, with legal services, miscellaneous services (which include data processing services), real estate, insurance agents, miscellaneous business services, banking, non-profit organizations and insurance underwriters also importantly involved. The most important manufacturing industries loading on this factor were printing, which has input-output ties to the service industries and non-basic manufacturing industries such as paints, bakery goods and cement. Some of the heavy manufacturing industries, among them blast furnaces and basic steel products, motor vehicles and iron foundries showed negative loadings on the factor, indicating that they were repelled by concentrations of employment in the higher order service industries.

Thus, the most important single grouping of industries was not a cluster of industries producing heavy or perishable goods, but a cluster of service industries. As one would expect from the nature of this factor, the city with the highest factor score was New York. Other cities scoring high were the very largest metropolises, such as Boston, Chicago, and San Francisco and smaller cities which are service centers for surrounding non-urban regions, such as Charlotte, N.C. and Des Moines, Iowa. As our research proceeds, we hope to test a number of hypotheses about why these service industries cluster and why the service cluster locates as it does.

A second factor was a rather close representation of a classic industrial complex -- the metal working and machinery complex. The industries and their factor loadings were:

Industry	Factor loading (Variable's correlation with factor)
Screw machine products and bolts	.907
Metalworking machinery and equipment	.797
Cutlery, hand tools and hardware	.786
Industrial leather belting	.586
General industrial machinery	.555
Mechanical measuring devices	.476
Metal cans	.487
Special industry machinery	.363

Neither primary steel nor iron and steel foundries had a strong connection with this factor. This replicated the finding of an Italian study that "the presence of a new iron and steel centre does not automatically give rise to new industries but only to isolated units such as cement manufacture from slag, etc. As for the attraction of an iron and steelworks to mechanical engineering industries, we have seen these tend basically to be set up in areas where the intermediate and auxiliary industries indispensable to them already exist. Modern conditions of efficiency and competitiveness, as may be relied upon by industries in the vicinity of the necessary auxiliary and subsidiary units largely offset the additional cost of transport of iron and steel products ..." (EEC, 1966).

Many of the specialized services which the other machine building industries require -- the making of jigs, molds and dies, and rebuilding of machine tools -- are included in the "metalworking machinery and equipment" industry. Other services, such as machine shops and electroplating also are in industries which had a relatively high loading on this factor. As we develop more disaggregated data on employment in non-manufacturing industries, we will be able to investigate the relationship of employment in engineering firms and research organizations to employment in machine building.

Among the places with the highest factor scores on this factor were Rockford, Ill.; Hartford, Conn.; Lorain-Elyria, Ohio; and Flint, Mich. Of the ten places scoring highest, all were within the so-called "Manufacturing Belt" in the Middle Atlantic and East North Central regions of the country. Thus while iron and steel were not heavily produced in places where machine building was important, production took place within easy rail (and in some cases water) distance of the important steel centers, and of such important consumers of machinery as the automobile industry.

Wages in the industries composing this cluster average about 7 percent above the mean for manufacturing as a whole. The impact of these higher wages and their impact on other sectors of the local economy is illustrated by the fact that manufacturing wages in the ten SMSAs scoring highest on this factor are 12 percent higher than the national average.

If the "metalworking factor" was an example of how a favorable industry mix can raise local wages, a "low wage/textile" factor showed the opposite kind of orientation. The industries composing this factor and their loadings were:

Industry	Factor loading (Variable's correlation with factor)
Cigars	.850
Female outerwear	.686
Pens, pencils, office supplies	.674
Children's outerwear	.399
Miscellaneous apparel	.275
Male furnishings	.269

These industries have wages more than thirty percent below the national average for manufacturing. Characterized by low skill and a high labor input, they are particularly important in such cities as Wilkes-Barre, York and Scranton, Pennsylvania; Lima, Ohio; and Tampa, Florida.

At times, the characteristic industrial structure of a handful of cities, or even of a single city makes them dominate a particular factor. For example, Reno and Las Vegas, Nevada have four times more per capita employment than any other city in industries in a "hotel-entertainment services" cluster. Similarly, the concentration of employment in fur goods, advertising, and small leather goods in New York probably accounts for the formation of a 'cluster' of these industries.

The distribution of factor scores can be used to derive a profile of the industrial structure of the SMSAs. We plan to compare factor score profiles among cities of different sizes and in different regions and to investigate how variations in industrial structure affect levels of income and cyclical stability.

The product moment correlation which we have used as the basis of our factor analysis is only one measure of industrial association. It assumes, perhaps unrealistically, that employment in a given industry varies linearly with employment in those industries with which it agglomerates. In fact, there may be important threshold effects or there may be scale economies which proceed at differing rates for each industry. These would lead to non-linear or perhaps even to discontinuous relationships. We are currently experimenting with a number of alternative measures. Rank order correlations compensate in part for non-linearities and reduce the influence of outlying observations. Discarding observations for which there is no employment in either of a pair of industries tends to reduce the extreme skewness in the frequency distribution of the data and enables us to study the conditional probability of having employment in one of a pair of industries, given that the other is present. We will also use the Automatic Interaction Detector program, which requires no prior assumption as to functional form, to investigate non-linearities in relationships. Although there are advantages to using relative or per capita employment in our analysis, this procedure may obscure important threshold or minimum scale effects. We are experimenting with absolute employment and similar measures in order to determine the sensitivity of our results to the method of measurement.

V. References

1. Alexandersson, Gunnar, The Industrial Structure of American Cities (Lincoln: University of Nebraska Press, 1956).

2. Alonso, William, "The Economics of Urban Size," Working Paper 138, (Berkeley: University of California, 1970). Also appearing in Papers of The Regional Science Association, Vol. XXVI, 1970.

3. Berry, Brian and Neils, Elaine, "Location, Size and Shape of Cities as Influenced by Environmental Factors: The Urban Environment Writ Large," in Harvey Perloff (ed.), The Quality of The Urban Environment: Essays on 'New Resources' in an Urban Age.

4. Borts, George and Stein, Jerome, Economic Growth in a Free Market (New York: Columbia University Press, 1964).

5. Chinitz, Benjamin, "Contrasts in Agglomeration: New York and Pittsburgh," American Economic Review, Papers and Proceedings (May 1961) pp. 279-289.

6. Dunn, Edgar S., Jr., "A Flow Network Image of Urban Structures," Urban Studies, Vol. 7, No. 3 (1970) pp. 239-258.

7. European Economic Community, Study for the Promotion of an Industrial Development Pole in Southern Italy (Brussels: European Economic Community, 1966). Economic and financial affairs series no. 5.

8. Florence, P. Sargant, et al., Industrial Location and National Resources (Washington: National Resources Planning Board, 1943).

9. Florence, P. Sargant, "Economic Efficiency in the Metropolis," in R.M. Fisher (ed.), The Metropolis in Modern Life (Garden City: Doubleday, 1955).

10. Fuchs, Victor, Changes in the Location of Manufacturing in the United States Since 1929 (New Haven: Yale University Press, 1962).

11. Hoover, Edgar M., Location Theory and the Shoe and Leather Industries (Cambridge: Harvard University Press, 1937).

12. Hoover, Edgar M., The Location of Economic Activity (New York: McGraw-Hill, 1949).

13. Hoover, Edgar M. and Vernon, Raymond, Anatomy of a Metropolis (New York: Anchor Books, 1962).

14. Isard, Walter, Location and Space Economy (Cambridge: MIT Press, 1956).

15. Isard, Walter, Schooler, Eugene, and Vietorisz, Thomas, Industrial Complex Analysis and Regional Development (New York: John Wiley and Sons, 1959).

16. Isard, Walter, Methods of Regional Analysis: An Introduction to Regional Science (Cambridge: MIT Press, 1960).

17. Isard, Walter, General Theory: Social, Political, Economic and Regional (Cambridge: MIT Press, 1969).

18. Jacobs, Jane, The Economy of Cities (New York: Random House, 1969).

19. Klaassen, Leo H., Growth Poles in Economic Theory and Policy, United Nations Research Institute for Social Development (UNRISD), March 1969.

20. Lichtenberg, Robert M., One-Tenth of a Nation (Cambridge: Harvard University Press, 1960).

21. McCarty, Harold, Hook, John and Knos, Duane, The Measurement of Association in Industrial Geography (Iowa City: State University of Iowa, 1956).

22. Meier, Richard L., A Communications Theory of Urban Growth (Cambridge: MIT Press, 1970).

23. Murphy, Raymond, The American City: An Urban Geography (New York: McGraw-Hill, 1966).

24. Nelson, Howard J., "A Service Classification of American Cities," Economic Geography, Vol. 31 (1955) pp. 189-210.

25. Ohlin, Bertil, Interregional and International Trade (Cambridge: Harvard University Press, 1933).

26. Perloff, Harvey S., et al., Regions, Resources and Economic Growth (Lincoln: University of Nebraska Press, 1960).

27. Richter, Charles E., "The Impact of Industrial Linkages on Geographic Association," Journal of Regional Science, Vol. 9, No. 1 (1969) pp. 19-28.

28. Robinson, E.A.G., The Structure of Competitive Industry (New York: 1932).

29. Rocca, Carlos A., "Productivity in Brazilian Manufacturing," in Joel Bergsman, Brazil: Industrialization and Trade Policies (London: Oxford University Press, 1970) Appendix 2.

30. Shefer, Daniel, Returns to Scale and Elasticities of Substitution by Size of Establishment for Two-Digit U.S. Manufacturing Industries, 1958, 1963. Regional Science Research Institute, Discussion Paper No. 26.

1 June 1971

31. Sonquist, John A. and Morgan, James N., The Detection of Interaction Effects: A Report on a Computer Program for the Selection of Optimal Combinations of Explanatory Variables (Ann Arbor: University of Michigan Institute for Social Research, 1964).

32. Stanback, Thomas M. and Knight, Richard V., The Metropolitan Economy (New York: Columbia University Press, 1970).

33. Streit, M.E., "Spatial Associations and Economic Linkages between Industries," Journal of Regional Science, Vol. 9, No. 2, (1969) pp. 177-188.

34. Thompson, Wilbur R., "The Future of the Detroit Metropolitan Area," in William Haber, et al., Michigan in the 1970's: An Economic Forecast (Ann Arbor: University of Michigan Graduate School of Business Administration, 1965).

35. Thompson, Wilbur, "Internal and External Factors in the Development of Urban Economies," in Harvey Perloff and Lowdon Wingo (eds.) Issues in Urban Economics (Baltimore: Johns Hopkins Press, 1968).

36. Ullman, Edward L., "Amenities as a Factor in Regional Growth," The Geographical Review, Vol. 44, No. 1 (January, 1954),pp. 120-132.

37. Vernon, Raymond, Metropolis, 1985: An Interpretation of the Results of The New York Metropolitan Region Study (Cambridge: Harvard University Press, 1960).

38. Weber, Alfred, Theory of the Location of Industries, tr. Carl Friedrich (Chicago: University of Chicago Press, 1929).

May 26, 1971

WELFARE ASPECTS OF NATIONAL POLICY

TOWARD CITY SIZES

Edwin S. Mills*

Note: This paper has been prepared for the Resources for the Future-University of Glasgow Conference on Economic Research Relevant to National Urban Development Strategies, Glasgow, Scotland, August 30-September 3, 1971. It is subject to revision and should not be cited or quoted without express permission of the author.

*Professor of Economics and Public Affairs, Princeton University, Princeton, New Jersey.

Edwin S. Mills was born in Collingswood, New Jersey in 1928. Although Collingswood is now part of the Philadelphia metropolitan area, it had the characteristics of a small town during his youth. The house in which he lived during his formative years was recently demolished to make a parking lot to serve the high speed rail line that now connects Collingswood with Philadelphia. After graduating from Collingswood high school in 1946, he served two years in the U.S. Army. He received his B.A. from Brown University in 1951 and his Ph.D. from the University of Birmingham in 1956. He was Assistant Lecturer at the University College of North Staffordshire 1953-55, Instructor at the Massachusetts Institute of Technology 1955-57, and held professional ranks at Johns Hopkins University from 1957 to 1970. He is now Professor of Economics and Public Affairs and Gerald L. Phillippe Professor of Urban Studies at Princeton University.

During recent years there has been a rapid increase in concern among scholars, public officials and others in the United States about the size and growth of the country's large cities, and it is now the stated policy of the national administration to discourage further growth of the largest cities.[1] In the United States, as in many other countries, there has been a long history of concern with allegedly excessive urbanization and the resulting drain of population from rural areas. In many other countries, including the United Kingdom, there is a history of concern and policy about growth of the largest cities. But serious concern with the size and growth of the largest cities is new in the United States, at least within living memory. This paper will examine the appropriateness of city sizes as a subject of public policy.

Many writers on urban affairs distinguish between urbanization and primacy. Urbanization refers to the number and percentage of people living in urban rather than rural areas. Primacy refers to the number and percentage of people living in the largest urban area. It is logically possible for either, neither or both urbanization or primacy to be excessive or deficient. Although I believe that much of the analysis in this paper applies to urbanization, it will not be discussed here. The paper is, however, closely related to the concept of primacy. Many Americans are especially concerned about the size of the New York metropolitan area, but the concern identified in the first paragraph usually refers to an unspecified number of the largest metropolitan areas rather than just to the largest area. Writers on the subject are not specific as to how many of the largest cities they think are too big. Nevertheless, it is clear that many people think the number is substantial. A guess is that much of the concern about excessive sizes of large cities in the United States includes the largest five or ten metropolitan areas. To fix ideas, according to the 1970 Census, New York is the largest metropolitan area with 16.0 million people, whereas St. Louis is the tenth largest with 2.3 million people. The exact number of cities to be included is not important for the discussion in this paper, the subject of which is the notion that a group of the largest metropolitan areas is larger than desirable.

It is clear that the appropriate objects of discussion are metropolitan areas rather than legal central cities. In the United States, central city boundaries are mainly set by historical accident and central cities now contain only about 45 percent of the population of metropolitan areas. The metropolitan area is the generic city, or at least the best measure that is available for a wide variety of data, and the term city will refer to the metropolitan area in this paper. When the legal city is intended, that term will be used. If the subject of concern were the legal city, the appropriate policy would be simply to move legal city boundaries to the optimum location. Although moving local government boundaries is not politically easy in the United States, it is a much less drastic policy than public intervention in private decisions about where to live and produce. But these are the policies proposed by those concerned with excessive city size, and the discussion only makes sense if metropolitan areas are the subject.

The final preliminary comment is that the observations in this paper are intended to apply only to the United States. Appropriate policy toward city sizes depends on many details of a society's institutional framework. Although I believe my assumptions are appropriate in the United States context, I have no views as to their applicability in the United Kingdom or elsewhere.

Some data on city size distributions

It has long been recognized that city sizes vary enormously within a region or country. The most superficial observation shows that the distribution of city sizes is strongly skewed to the right, i.e., there are many small cities and a few very large cities, in almost all countries and at almost all times. Any public policy toward city sizes is an attempt to alter the size distribution of cities.

There is of course a long history of empirical studies of city size distributions, going back to the 19th century.[2] At one time, most writers on the subject believed that the distribution conformed to the rank-size rule in almost all countries and all times. The rank-size rule is the distribution in which a city's population is inversely proportional to its rank. I believe that most careful students of the subject now agree that most data are about equally well described by any member of a class of skewed distributions that includes the Pareto and the log-normal. Furthermore, either the form or the parameters of the distribution vary from country to country. For example, Davis[3] recently computed a measure of primacy for all the countries in the world that had at least four cities of at least 100,000 population each in 1960. The measure is the population of the largest city divided by the sum of the populations of the next three largest cities. For the rank-size rule, this statistic has a value of 12/13. Among the 46 countries included in Davis' tabulation, the primacy statistic varied from a low of 0.51 to a high of 4.64. It is thus clear that deviations from the rank-size rule are substantial.

The point of the last paragraph is that the evidence suggests that the size distribution of cities is not immutably determined by technical or other conditions and that public policy could presumably affect it if it was desirable to do so. However, the evidence also suggests that city size distributions within countries are remarkably stable over long periods of history. In fact, I would say that the persistence of particular city size distributions through time is one of the most remarkable regularities in the social sciences.

The evidence and literature on the stability of city size distributions suggests that public policies can alter the distribution, but that the task is likely to be very difficult. Measurable changes in the distribution within a period of a decade or two are likely to require massive public intervention in location and land use decisions. Relatively minor policy changes such as residency requirements for public welfare recipients, diversion of federal grants from large city governments or locating a few federal facilities in small cities or towns are unlikely to have noticeable effects. I believe my view is confirmed by postwar attempts of European governments to curtail the growth of large cities.

My final observation in this brief survey of city size distributions is that the United States does not have a strong concentration of its population in its largest cities in comparison with other industrialized countries. Davis' 46 countries have an average primacy index of 1.42, whereas the United States index is only 0.88. France, West Germany, the Soviet Union, the United Kingdom and Japan have higher indices. The New York metropolitan area contains only eight percent of the country's population, a smaller percentage than is found in the largest metropolitan area in most industrialized and many other countries. The ten largest metropolitan areas in the United States contain only about 30 percent of the country's population, a smaller share than in most countries. Of course, these data do not refute the contention that the largest United States cities are too big, but they certainly do not lend it any support.

Determinants of city size distributions

A fundamental way to characterize the purpose of a city is to say that it is to facilitate production and exchange of goods and services by proximate locations of diverse activities. If one includes public and private non-profit production of goods and services, the characterization applies to cities that are seats of government or of religious activity. But in the United States, with the sole exception of Washington, D.C., locational decisions of households and profit-making firms have been dominant in the growth of large cities. The following discussion will therefore focus on the household and business sectors of the economy.

A highly specialized economy requires not only the concentration of substantial amounts of production in one location, but also large numbers of exchanges in the production and distribution of goods and services. Exchanges require movement of goods and people as well as communication between people. Movement, or transportation, and communication require the use of valuable resources, including the time of people. Cities economize on the use of valuable resources in transportation and communication by locating large numbers of related activities close to each other or, what is the same thing, by production of goods and services with very high ratios of capital and other inputs to land in a relatively small area. The most important and widely studied urban exchange is that of labor services for wages and salaries, and the related transportation, or commuting, of workers. But it is of course only one of many exchanges that take place in large cities.

Manufacturing is of course the classical example of specialization in economic activity and the close historical relationship between urbanization and the growth of manufacturing is well known. But in recent decades urbanization has been only loosely related to industrialization. In the United States, for example, except for cyclical movements, about 25 percent of the labor force has been engaged in manufacturing since about 1920, whereas the urban population has grown from about half to nearly three-quarters of the total during that half century. The reason is that the rapidly growing sectors, mostly services, are highly specialized and indeed most are more concentrated in large metropolitan areas than is manufacturing. The phenomenon is also observed in underdeveloped countries, most of which are urbanizing rapidly. A rather foolish literature has appeared decrying overurbanization of underdeveloped countries in which urbanization is outdistancing industrialization.[4]

The idea that cities economize on the use of resources by concentrating specialized activities forms the basis of the only economic theory of the size distribution of cities that we have. Beckmann,[5] building on the work of Lösch and others, assumed that natural resources are uniformly distributed, that in each activity a worker can supply the needs of a fixed number of consumers, and that a city of given size can supply the needs of a fixed number of consumers, and that a city of given size can supply the needs of a fixed number of cities of the next smaller size. In part, these assumptions rely on Lösch's model of the spatial distribution of cities for their justification. Beckmann showed that these assumptions imply a size distribution of cities that can be approximated by the rank-size or Pareto distributions. Although the Lösch-Beckmann model can hardly be regarded as realistic in any detailed sense, it is remarkable that a distribution of city sizes that is at least approximately realistic can be deduced from such a simple model. I believe it is one of the most important insights in urban economics.

Many things that economists rightly think to be important are excluded from the Lösch-Beckmann model. It has no demand side and only the most primitive production functions. Most important, it ignores the uneven distribution of natural resources, which leads to regional comparative advantage. International trade specialists usually assume that international comparative advantage depends on differences among nations in endowments of physical and human capital, technology and natural resources. Human resources, capital and technology are much more mobile among regions of a country than among countries and can be ignored as sources of regional comparative advantage at least as a first approximation. But natural resources, including minerals, topography and climate, vary greatly among regions of the United States and certainly have major effects on city sizes. Indeed, the factors that cause international comparative advantage affect the locations and sizes of cities within a country. Even if ~~natural~~

natural resources were uniformly distributed within the United States, city sizes and locations would be affected if international comparative advantages induced trade with the rest of the world. It is a remarkable fact that most large cities in the United States are located with easy access to the sea despite the fact that only a small proportion of output enters international trade. Not only the cities on the Atlantic, Gulf and Pacific coasts are in this category, but also those on the Great Lakes.

What is the effect of regional comparative advantage on the distribution of city sizes? What modifications are required in the Lösch-Beckmann model? It seems reasonably clear that regional and international comparative advantage has resulted in a quite different spatial distribution of cities in the United States from that predicted by the Lösch model. The Lösch model would certainly not predict the concentrations of population found around the borders of the country. However, it is much less clear what modifications comparative advantage entails for the size distribution of cities. No theoretical work has appeared and it may be that none can be done except within the context of very detailed characteristics of particular countries. Some ad hoc observations are easy to make. Presumably, if New York were the only good natural harbor on the east coast, it would occupy a more dominant position among United States cities. Or, if the climate were less attractive in Southern California than it is, our second largest metropolitan area would be smaller relative to the largest than it is. It would be interesting to have a simple characterization of regional comparative advantage that would tell us whether the actual size distribution of cities should be more or less skewed than that predicted by the Lösch-Beckmann model, but I know of no such work.

Before concluding this section it should be mentioned that there is a statistical theory of city size distributions that was first put forward by Simon.[6] It has long been known that a random variable will be approximately normally distributed if it is a weighted sum of a sufficiently large number of well-behaved random variables. It follows easily that if a random variable is the product of a sufficiently large number of well-behaved random variables, each raised to a power, then its logarithm will be approximately normally distributed. Simon showed that this process, called the law of proportionate effect, along with assumptions about the birth-death process, leads to a class of skewed distributions of which the log-normal and Pareto are special cases and generally good approximations. In the case of city sizes, the birth-death process means the process by which communities attain or fall short of the population needed to be counted as cities. It is easy to think of phenomena that affect city sizes in proportionate rather than additive fashion. For example, a change in the overall population's birth or death rates is likely to affect the population of each city by an amount that is about proportionate to the city's population.

Although the law of proportionate effect is a statistical rather than an economic theory, it is not independent of economic assumptions. The law of proportionate effect has been used as an interesting explanation of the highly skewed distribution of firm sizes. In that context, it is a plausible explanation only if firms' long run average cost functions are flat. Some such assumption must also be implicit in the application to city sizes. If there were important diseconomies to large cities then they, rather than the law of proportionate effect, would determine city sizes.

Welfare economics and city size distributions

It is usual to classify welfare considerations into efficiency and equity aspects. Although it is possible that the distribution of wealth or income is related to the distribution of city sizes, it is unlikely that the relationship is at all close. If it is not, then the distribution of city sizes can be evaluated by the usual efficiency criteria of welfare economics.

The usual procedure in welfare economics is to deduce a market allocation of resources from assumptions about technology, tastes and factor availability, to ascertain whether the market allocation satisfies conditions for Pareto-efficiency, and to propose policies to correct any inefficiency that is found. Unfortunately, the one economic model of city size distributions available, the Lösch Beckmann model, is not rich enough for the purpose. Technological, taste and factor availability assumptions are not sufficienty articulated to be able to distinguish between market and efficient distributions of city sizes. In the absence of appropriate models for overall welfare analysis, it is necessary to lower our sights somewhat and ask what specific reasons there are for thinking that private choices might lead to at least some cities that are too large.

A preliminary observation is that a remarkably large number of economists and other social scientists have speculated about _the_ optimum size of a city.[7] If the analysis in the previous section has any validity at all, the notion of a single optimum city size is absurd. I am unable to imagine what a model might look like in which all cities in a country had the same optimum size. It seems inconceivable that it would bear any relationship to the real world.

Serious allegations about excessive city sizes in the United States can usually be included under congestion, pollution and the operation of the public sector. In each case the argument is that people or firms that decide to locate in large cities bear only part of the cost that their decisions impose on the urban residents as a whole. The part that is not borne by the decision maker is an external diseconomy and is subject to the usual welfare analysis. In an important recent paper, George Tolley[8] has shown that the resulting resource misallocation may entail cities that are too large or too small, depending on the details of the model. The rest of this section presents more detailed comments on the three kinds of externalities mentioned above.

Congestion is the best understood. Time costs are an important part of the total cost of urban travel and the time cost per trip is an increasing function of the number of travelers using the transportation system, at least beyond the usage rate at which congestion begins. The implication is that each additional traveler slows other travelers. The cost he thereby imposes on them would be difficult for him to take into account and he has no incentive to do so. The result is that the urban transportation system is excessively congested and travel costs are higher than optimum. Economists and others have proposed congestion tolls and other means of optimizing the system, and cities have experimented with a variety of methods of rationing the use of streets and highways.

Although the analysis is almost certainly correct, it would be nice to know more about its quantitative importance. Some people write as though an optimum urban transporation system would have no congestion, but it is clearly wrong. The phenomena discussed earlier that justify extremely intensive use of urban land apply as much to land used for transportation as to land used for buildings. It is the fact that they make intensive use of land that justifies mass transit systems in large cities. Within a mode, the way to use land intensively is to have a large ratio of travelers to lane miles of facility, i.e. to congest the facility. Mills and de Ferranti[9] have shown how to compute optimum congestion in a simple theoretical model.

The more important point is that the distorting effects of excessive congestion almost certainly have to do with the structure rather than with the size of the metropolitan area. Congestion results almost entirely from concentration of activity near the centers of cities, rather than from the overall size of the metropolitan area. It is always possible to eliminate congestion by decentralizing economic activity in a metropolitan area. It seems likely that the main effect of congestion tolls that reduced the use of streets in large United States cities would be to speed up the process of employment suburbanization that has been so rapid since World War II.

I believe that the point in the previous paragraph is crucial to all the arguments for controlling city sizes. To the extent that urban markets distort resource allocation, I believe that distortion mainly affects the structure rather than the size of metropolitan areas. The appropriate public policies are those that improve the incentive system, such as congestion tolls. Policies to control the size of the metropolitan area are as likely to be in the wrong direction as in the right direction, and are inefficient ways of dealing with the problems even if they are in the right direction. Thus, I believe that public policy should be concerned to improve market incentives rather than to control city sizes.

Having gone through the analysis with congestion, pollution can be dealt with briefly. The basic nature of the external diseconomy is similar. Discharges of wastes to the air or to bodies of water are economical to the firm, household or public agency responsible for the discharge, but impose costs on users of the environment which the discharger lacks incentive and ability to take into account. The costs so imposed on users of the environment are thus external diseconomies and justify public intervention. The most obvious kind of intervention is a charge on harmful discharges to the environment, or effluent fee, which is analogous to a congestion toll. One response to an effluent fee, or to regulation of discharges, is to halt the discharging activity or to move it outside the metropolitan area. But it is hard to imagine public policies in the United States that knowingly had these effects. More likely, and more desirable, responses are changes to less polluting productive processes and the collection and treatment of harmful wastes. Although these responses improve resource allocation, they do not reduce the sizes of large cities. Indeed, in most waste treatment systems, collection costs are a large part of the total, and collection costs are smallest in concentrated areas like large cities. Thus, to the extent that increased waste treatment is the response to public anti-pollution policies, it may increase the sizes of large cities. Once again, the conclusion is that public policies to reduce city sizes may well do more harm than good, and are at best inefficient ways of dealing with the problem.

A third reason that some people believe makes the largest United States cities too large is the operation of the public sector.[10] A characteristic of the federal system of government in the United States is that increasing shares of local government expenditures are financed by grants from state and federal governments. Some people believe that the provision of funds by state and federal governments to finance services and transfers to low income residents in central cities provides poor people with excessive incentive to move to central cities of large metropolitan areas from small towns and rural areas.

Some facts will suggest the likely importance of the phenomenon. In 1964-65, state and federal grants to local governments in the central cities of the 37 largest metropolitan areas in the United States were $88 per capita.[11] In the suburbs of the 37 metropolitan areas, the figure was $80. For the remainder of the nation, it was $76. These data suggest that the incentive resulting from outside financing of local government expenditures for a poor family to migrate to a metropolitan central city is at most $12 per capita per year. There are problems with these data, but even if the correct figure is several times $12, it is nevertheless more than offset by differences between tax levels. State and local taxes in the same 37 central cities were $200 per capita. In the suburbs they were $152 and in the remainder of the nation they were $103. Thus, state and local taxes were about $100 more in central cities than in areas from which poor people might migrate. It is extremely unlikely that the "subsidy" to central city governemnts from state and federal governments has a substantial effect on migration.

Although subsidization of central city residents by higher levels of government is unlikely to be important, availability of public services and transfers may be. The issue is related to racial problems in the United States. Black Americans receive, and pay for, much better education and other public services in metropolitan central cities than they could obtain in the rural south from which they have migrated in large numbers. Likewise, they were frequently deprived of public assistance for which they were legally eligible in the rural south, whereas they have access to public assistance in northern cities. Deprivation of public services and transfers was part of the terrible oppression to which black Americans were subjected in the rural south. Nobody knows to what extent better access to public services and the political process has induced blacks to migrate. Most studies suggest that employment opportunities provide the primary incentive. But it is not inconceivable that the availability of public services and transfers has been of significant importance. If so, the implication for public policy is to enable blacks to achieve the same access to the political process in the south that their numbers and concentration have enabled them to achieve in northern cities. It is possible that the result would be some education in the migration of blacks from the rural south to northern cities, although I doubt that the effect would be substantial. But the effect on migration and city size is incidental. The appropriate policies are those that correct the defects in the political system. Indeed, to impose barriers to their migration to northern cities would condemn them to the hostile environment which they have fled in such desperation.

I conclude that proposals to discourage the growth of large cities in the United States are misguided. Appropriate public policies should be aimed at the specific reasons for resource misallocation. The effects of such policies on city sizes would be incidental and mostly unpredictable.

Some observations on the urban crisis in the United States

If, as I believe, the case for public policies to curtail the growth of large cities is weak, how is one to account for the widespread belief in their desirability among public officials and others?

The most obvious reason for the widespread bias against large cities, which extends far beyond the United States, is that the disadvantages of large cities are obvious, whereas the advantages are subtle. Congestion and pollution are apparent to all, whereas the efficiency of production and exchange which cities permit are abstract concepts. Moreover, public officials are painfully aware of the turmoil in cities and of the organized demands for political participation engendered by city life.

In the United States, however, it seems clear that concern about city sizes is part of the sense of an impending urban crisis whose causes go far beyond the phenomenon described in the previous paragraph. Newspapers and magazines give the impression that American central cities are sinking rapidly and will soon disappear from sight. Mayors of large cities freely predict imminent bankruptcy of their governments. Even social scientists, who should know better, write articles that suggest that everything is getting worse and worse in the cities.

The malaise about the plight of the cities is concentrated on the legal cities and especially on conditions of life in the black ghettoes of central cities. It is well known that blacks have migrated in enormous numbers from the rural south to segregated parts of northern central cities since World War II. Blacks have urbanized in proportionately larger numbers than whites, and the black population is now, for the first time in history, more urbanized than the white. The paradoxical nature of the sense of crisis is that the data show clearly that massive urbanization has been a major factor in improving conditions of life for blacks. Whether one looks at data on income, poverty, housing, education, health, mortality or political particiaption, the story is the same.[12] Conditions of life are much better for black Americans in metropolitan areas than in the small towns and rural areas from which they have migrated and improvements have been much more rapid. Almost the sole exception is crime statistics, which show that crime, especially against blacks, is worse in big cities than elsewhere.[13] Although crime is certainly a serious and growing problem in the United States, meaningful comparison is not possible because crimes against black people were largely unrecorded in the rural south.

There is really nothing very surprising about all this. American cities are again playing the role of absorbing and improving the lives of millions of immigrants that they played in the 19th and early 20th centuries. The difference is that the post-World War II immigrants have come from the rural south instead of from abroad. From a historical perspective, the surprising aspect of the present situation is the sense of gloom that pervades popular writing about the cities.

Of course, every large migration to American cities has been accompanied by tension, conflict and trauma. Undoubtedly, the current trauma in cities is the worst of all because of powerful racial antagonisms. Conservatives are enraged at the prospect of blacks taking control of local governments as they attain numerical majority in many central cities during the 1970's. Liberals are depressed by the failure to achieve an integrated society.

Urbanization of the black population is one of the great dramas of American history. It provides enormous social problems and scope for social science research, all of which are beyond the scope of this paper. Its relevance to the subject of this paper can be stated simply. Public policies to curtail the growth of large metropolitan areas are extremely unlikely to be directed at growth of middle class white suburbs by the present or any likely national administration. Thus, any such policies will inevitably be directed at migration of poor blacks to central cities. These policies would inevitably cut off the most important means by which blacks have been able to escape oppression and improve their lives during the last three decades.

FOOTNOTES

1. See the paper by Lowdon Wingo prepared for this conference for a thorough survey of recent literature.

2. See Berry, Brian and William Garrison, "Alternative Explanations of Urban Rank-Size Relationships," Annals, Association of American Geographers, V. 48, 1958, pp. 83-91.

3. Davis, Kingsley, World Population, 1950-1970. Population Monograph Series No. 4, Berkeley: University of California Press, 1970.

4. See Mills, Edwin. "City Sizes in Developing Economies," forthcoming.

5. Beckmann, Martin. "City Hierarchies and the Distribution of City Size." Economic Development and Cultural Change, V. 6, 1958, pp. 243-248, and references therein.

6. Simon, Herbert, "On a Class of Skew Distribution Functions," Biometrika, V. 42, 1958, pp. 425-440.

7. See the paper prepared by Wingo for this conference for references to the literature.

8. Tolley, George, "The Welfare Economics of City Bigness," unpublished.

9. Mills, Edwin S. and David de Ferranti, "Market Choices and Optimum City Size," American Economic Review, V. LXI, No. 2, May 1971, pp.

10. Several papers deal with this issue in Margolis, Julius (editor), The Analysis of Public Output. New York: Columbia University Press for the National Bureau of Economic Research, 1970.

11. The data in this paragraph are from Advisory Commission on Intergovernmental Relations, Fiscal Balance in the American Federal System, (A-31; October 1967), Vol. 2.

12. The data are readily available, and some have been collected together in Mills, Edwin, "An Optimistic View of the Urban Crisis," unpublished.

13. See the paper by Irving Hoch prepared for this conference.

May 25th 1971

The Pure Theory of City Size in an Industrial Economy

Alan W. Evans*

NOTE: This paper has been prepared for the Resources for the Future - University of Glasgow Conference on Economic Research Relevant to National Urban Development Strategies, Glasgow, Scotland, August 30-September 3, 1971. It is subject to revision and should not be cited or quoted without express permission of the author.

*Lecturer in Urban Studies

Department of Social and Economic Research

The University of Glasgow

Alan Evans qualified as a Chartered Accountant in 1961 before going to University College London to read Philosophy and Economics. After postgraduate work at the University of Michigan and University College he joined the Department of Social and Economic Research at the University of Glasgow in 1967 where he has worked on the economics of land use planning for three years and is now working on the economics of the size of cities. In October 1971 he is to take up a post at the Centre for Environmental Studies in London.

One of the most well known empirical regularities in urban studies is the existence within almost every national economy of a size distribution of cities and towns which is statistically fairly regular. Yet "no scientific explanation worthy of the name has been advanced so far" (Tinbergen(1968)). In this paper we develop a theory of city size which provides an economic explanation for the existence of a regular size distribution of cities in an industrial economy. We also show that the assumptions and the predictions of the theory are in accord with the empirical evidence.

The structure of the paper is as follows. We first review the two alternative approaches to the theory of city size which generate predictions of an urban hierarchy. Next we state, and attempt to justify, the assumptions we intend to make. In the third and fourth sections we show that both wage rates and rents increase with city size and in the fifth section we show that the cost of certain services can be expected to decrease with city size (the so-called 'external economies', or 'localisation economies' or 'agglomeration economies'). In the sixth section we consider the single firm and show how it selects its optimal location among the set of cities of varying sizes, whilst in the seventh section we show that in general equilibrium the location decisions of all the individual firms acting separately will cause the existence of an urban hierarchy.[1] Some of the welfare implications of the theory are discussed in the eighth section and in the ninth section we discuss the effects of a reduction in the cost or an increase in the speed of commuting on the structure of the urban hierarchy. Finally we discuss the relationship between the theory presented in this paper and other theories of location and suggest ways in which the theory might usefully be extended.

May 25th 1971

I Alternative Theories of the Urban Hierarchy

The main attempts to explain the empirical regularities in the size distribution of cities and towns fall into two main categories, systems theory and central place theory.

(1) Systems Theory (See, for example, Simon (1955), Berry (1964), Berry and Woldenberg (1967)). An approach using systems theory is to explain the empirical regularity as being the most probable statistical distribution under certain conditions. In particular it may be assumed that the rate of growth or decline in the size of any city is uncorrelated with its size. Although if viewed purely as applied mathematics these theories are intellectually stimulating, in most respects they are not very useful since the only thing they can 'explain' is the existence of an urban hierarchy. Furthermore, the theories rest on insecure foundations since the assumption mentioned above would seem to be wrong, at least in the short run, since recent work has shown that rates of growth are correlated with city size (Stanback and Knight (1970), von Böventer (1969), Muth (1968)).

(2) Central Place Theory (See, for example, Berry (1967), Berry and Pred (1961), Parr (1970)). Central place theory explains the existence of the urban hierarchy in terms of a hierarchy of market areas. The larger the city the wider the range of goods and services which it provides and the larger its market area.

Central place theory provides an explanation of the urban hierarchy which is more satisfactory than systems theory since it yields predictions which can be tested and, indeed, have been, extensively, see for example Berry (1967). Thus its explanation of the urban hierarchy is buttressed by empirical research relating to other aspects of the spatial economy whilst the explanation provided by systems theory must needs stand on its own. Nevertheless the empirical regularity in the size distribution of cities in an industrial economy cannot be completely explained by central place theory since it can only account for the spatial distribution of retail trades, services, &c. Whilst central place theory may be adequate to explain the urban hierarchy in areas such as S. W. Iowa where there is little manufacturing, it is inadequate to explain the size distribution of cities and towns in manufacturing areas such as the N. E. United States or the United Kingdom.[2] "The existence of urban hierarchies in the heavily industrialised regions remains a fact in search of a theory" (Higgs (1970)).

II The Assumptions of the Theory

In this paper we develop a pure theory of city size. We call it a *pure* theory not because we intend the theory to be understood as an abstraction having little to do with the real world but because we have tried to abstract completely from all aspects of the economy which we judge to be irrelevant. This has been done for two connected reasons. In the first place it is easier to carry out the analysis and in the second place it is easier to understand.

The assumptions we wish to make are these.

(1) There are two sorts of industry in the economy. We call these 'manufacturing' industry and 'business services' or, sometimes, services. These names are meant to have only a slight descriptive content.

(2) All the output of the firms providing business services is sold to firms in manufacturing industry.

(3) All the output of the firms in manufacturing industry is sold to consumers or to other firms in manufacturing industry.

(4) The cost of transport of the output of manufacturing industry is zero so that the sales of any manufacturing firm do not vary with its spatial location.

(5) The cost of transport of the output of the firms providing services is infinitely high but the whole range of business services is provided in every urban place. Since business services are an input to manufacturing firms, both business services and manufacturing firms locate in the centre of each city (the Central Business District or CBD). It might possibly be more realistic to assume that the cost of transport of business services is very high whilst all business services are not provided in every city. This would necessitate the introduction of some type of central place theory to explain the location of the services which are not available everywhere. I do not think the small gain in the explanatory capacity of the theory would be enough to compensate for the increase in complexity. We thus ignore what von Böventer (1970) calls the common agglomeration economies and positive overspill effects which arise when a town is located close to another (bigger) centre.

(6) The plants which provide business services are indivisible. There are therefore economies of scale in the provision of business services.

(7) All firms attempt to maximise profits and households attempt to maximise utility. We assume perfect competition between firms except that where this is not possible, in the case of some business services, we assume that the services are sold at average cost (including 'normal' profits).

(8) The cost of commuting within the city is neither zero nor infinite but is at some intermediate level. Since employment in any city is at a single workplace the analyses of residential location by Alonso (1964), Wingo (1961), or Muth (1969) are applicable. Furthermore we take over unchanged the simplifying assumptions which are made by Alonso in order to develop a workable theory of residential location, e.g. each city is situated on a flat plain with no topographical features, the cost and speed of commuting is uniform in all directions, &c.

(9) All commuters' fares are equal to marginal social cost. There is no unpriced congestion. In particular there is an efficient road pricing system. We thus avoid having to consider a problem which has tended to dominate at least one earlier discussion of the economics of city size (Neutze (1965)). See also Reynolds (1966).

(10) Households can remove themselves from one permanent location to another at negligible cost.

It can no doubt be argued that the above assumptions are unrealistic, and it must be admitted that to some extent they are. But they are no more unrealistic than the assumptions which are made in many other branches of economic theory. The closest case conceptually is the theory of international trade where assumptions such as 'perfectly competitive markets, the absence of transport costs, the complete mobility of factors within countries and immobility between countries are still made in the basic formulations of the modern theory' (Findlay (1970) p.17). The theory of international trade explains international trade in a non-spatial world economy. We explain the size distribution of cities in a (virtually) non-spatial national economy. In both theories it is hoped that the assumptions are realistic enough to derive useful explanations and predictions whilst being unrealistic enough to handle conceptually.

III <u>Input Prices as a Function of City Size - Rents</u>

We can classify the inputs of firms in manufacturing industry as labour, land (or floor space), capital or business services. In this section and the two following sections we shall show that, with one exception, the prices of these inputs can be expected to vary with city size. Hence manufacturing firms must take these price variations into account in selecting the city in which to locate. The exception is capital. There is no reason to suppose that the cost of capital should vary with city size nor is there any evidence that it does so.

We shall show in this section that rent, the price of floor space, will tend to increase with city size but at a diminishing rate.

Figure 1 shows a cross-section through the rent surface of a city of a given size. The rent per unit of floor area is indicated on the vertical axis and distance from the city centre on the horizontal axis. The city centre is

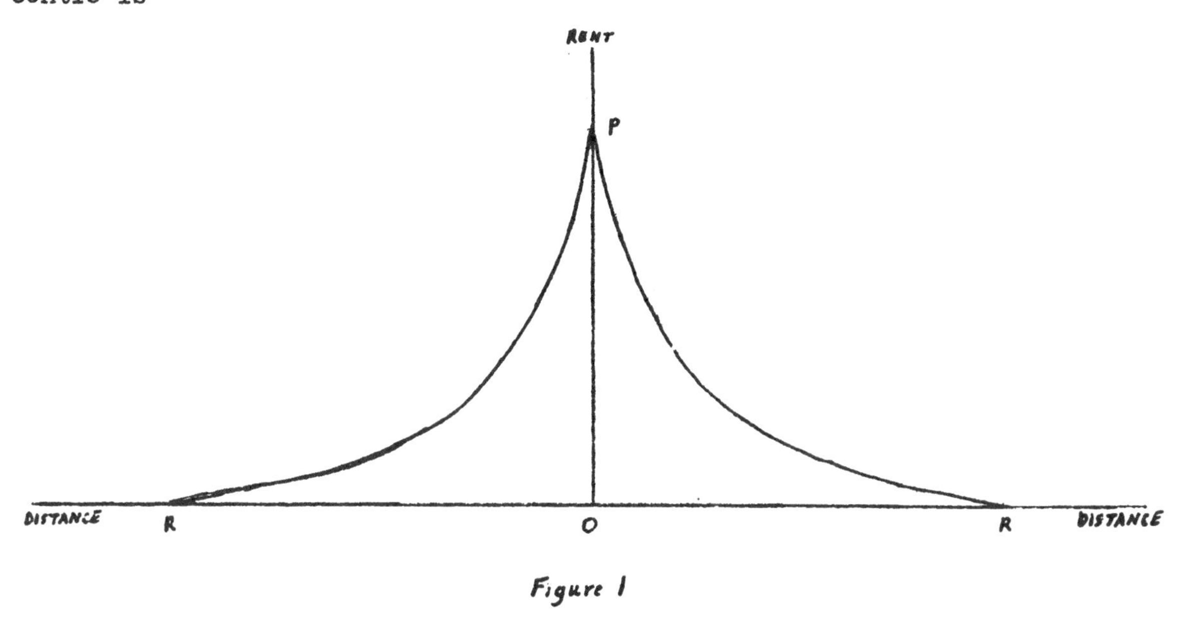

Figure 1

denoted by the point O , the radius of the built up area of the city is OR , and the rent at the city centre is OP. We assume that the market for space in the city is in long run equilibrium. The density of floor space (and persons per acre) is thus a monotonic function of the rent per unit of floor space, the characteristics of the function being determined by the state of building technology. Both rent and density are highest at the city centre and decline at a decreasing rate as distance from the city centre increases. On this see particularly Muth (1969) Chapter 3.

Now compare the long run equilibrium in this city with the long run equilibrium in a city which is substantially larger. Since the population of the large city is, by definition, larger either the population density must be greater at some distances from the centre or its built up area must be greater or both. But the density will be greater at some distance only if the rent is higher at that distance since, by assumption, both urban housing markets are in long run equilibrium. Provided that the characteristics of the populations of the two cities are similar there is no reason to suppose that the shape of the rent surface of the larger city will differ substantially from the shape of the rent surface of the smaller city. Hence if the rent at some given distance from the centre in the larger city is higher than it is at the same distance from the centre in the smaller city, it will be higher at all distances from the centre; if it is higher at all distances from the centre then the radius of the larger city will be greater than that of the smaller, and rents at the centre of the larger city will be higher than in the smaller.

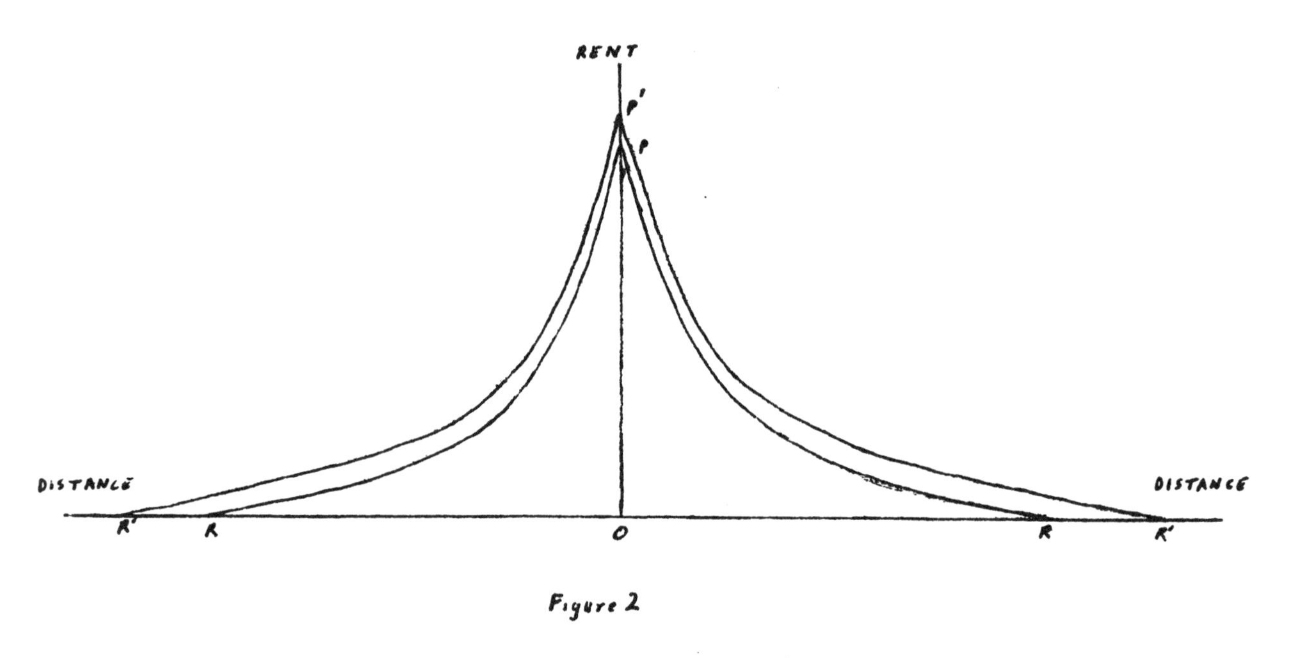

Figure 2

In Figure 2 the cross section through the rent surface of the larger city is shown by the curve R' P' R' which lies above the curve RPR at all distances from the centre. In particular OP' > OP; we would expect city centre rents to increase with city size.

Little direct evidence is available to support this assertion though casual empiricism would suggest that two corrollaries of the argument are correct. The density of floor space at the city centre (indicated by the height of buildings) tends to increase with city size, and the radius of the built up area of the city tends to increase with city size. So far as I am aware more direct evidence is only available for Switzerland where Widmer (1953) found that the rent of housing was positively correlated with city size.

We can only go on to assert that the rent of floor space at the city centre will not only increase with city size but will tend to increase _at a diminishing rate_. Figure 3 indicates the expected relationship. The rent per unit of floor area is shown on the vertical axis and city size (measured by population) is shown on the horizontal axis. The rate of increase in the

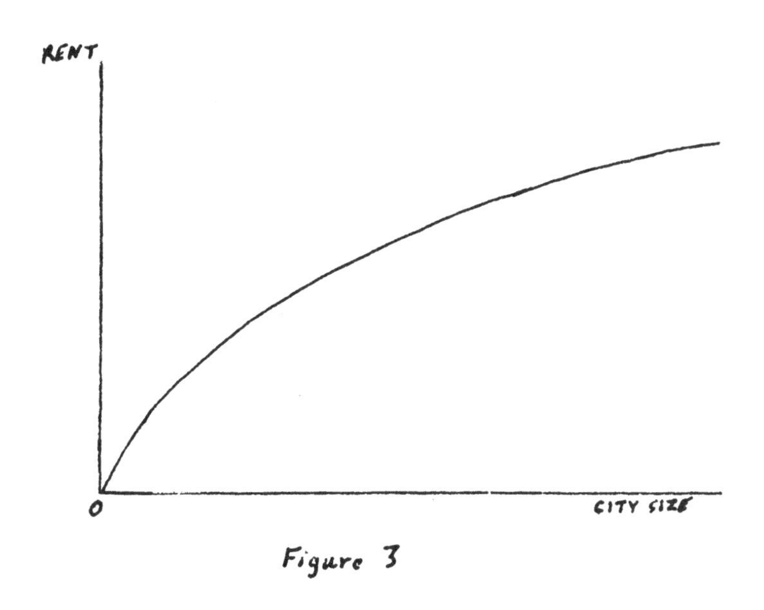

Figure 3

rent of space will decline with city size because the rent at the centre is proportional to the radius of the built up area of the city whilst the population of the city is proportional to the area and, hence, to the square of the radius. As the area of the city increases, a given increase in the population can be accommodated by smaller and smaller increases in the radius of the city and will therefore result in smaller and smaller increases in central city rent. As evidence in support of the assertion it may be noted that Widmer (1953) fits a semi-logarithmic linear regression to his data for Swiss cities and whilst he gives no statistical evidence to suggest that this form of regression fits the data better than a linear form, he does publish the scatter diagram which shows that the semi-logarithmic transformation fits the data well.

IV Input Prices as a Function of City Size - Wages

In this section we shall show that wage rates will tend to increase with city size at a decreasing rate. The argument is a continuation of that in the preceding section.

We use as a tool of analysis the concept of the set of bid price curves introduced by Alonso (1964). A bid price curve connects the set of points indicating rents and locations between which a household is indifferent. A household's preferences are described by an infinite set of bid price curves just as in the usual analysis of consumer behaviour a household's preferences are described by an infinite set of indifference curves. However whilst the household prefers to consume the bundles of goods represented by points on higher indifference curves to those represented by points on lower indifference curves, since it obviously prefersto pay a lower rent to a higher rent at any given distance from the centre, it must prefer locations and rents represented by points on lower bid price curves to those represented by points on higher curves. At its optimal location therefore the rent gradient will be tangential to the lowest attainable bid price curve from above. In Figure 4 we reproduce one half of Figure 1 but show in addition one of the bid price curves of a representative household.

Figure 4

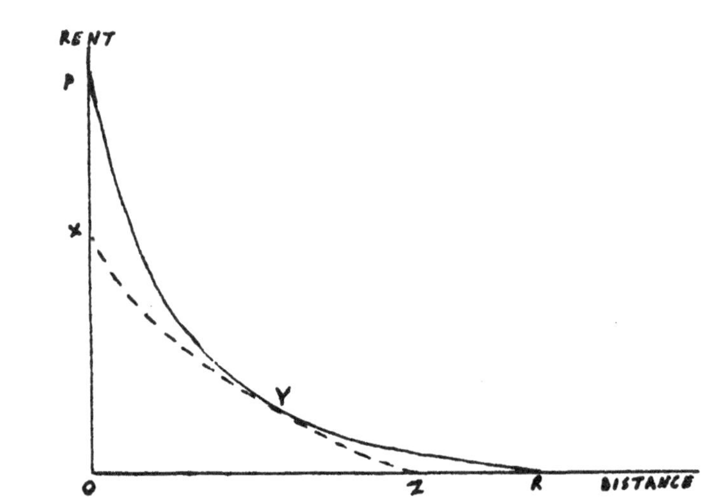

The curve XYZ is the lowest bid price curve attainable by the household; the point Y indicates the optimal location and rent. Suppose now that the household, with an unchanged schedule of preferences, is transferred to a larger city where the rent surface is that indicated by the curve R'P'R' in Figure 2. The bid price curve XYZ is no longer attainable since R'P'R' lies everywhere above RPR and therefore lies everywhere above XYZ. The lowest bid price curve which is now attainable by the household is one which is tangential to P'R' and this must be a higher bid price curve than XYZ. The household is made worse off by an increase in city size because of the increase in its rent and/or travel costs. If migration between cities is possible at a negligible cost, some benefit must accrue to residents of the larger city which compensates them for the higher rent and/or travel costs payable in the larger city and prevents them migrating to the smaller city. In the real world these compensatory benefits might accrue through cheaper food or other goods, or through the availability of certain services (e.g. theatres, art galleries, &c.) or merely through the ambiance of the larger city. There is however a significant amount of evidence to suggest that higher money wage rates provide at least a significant part of the compensatory benefit in the real world, and this is the only form of compensation which does not conflict with the assumptions of the theory.

Widmer (1953) found, using simple regression, that the average hourly rate of pay in Switzerland was positively correlated with city size. Fuchs (1967) concluded from his analysis of U.S. data using multiple regression that "the relation between earnings and city size is large, persistent, and cannot be explained by correlation between city size and other variables". The author found a generally significant positive correlation between British clerical salaries and city size (Evans (1970) Chapter XII)

In the same way that we showed that rents can be expected to increase at a decreasing rate, we can show that wage rates should not only increase with city size but should also increase at a decreasing rate. The population of the city is proportional to the area of the city and hence to the square of the radius, whilst the rent and travel costs are proportional to the radius of the city and hence to the square root of the area. Thus the increase in wage rates necessary to compensate for the increase in rent and travel costs consequent upon a given increase in the population of the city will get less and less as the city increases in size. As evidence in support of the assertion we may note that Widmer fits a semi-logarithmic linear regression to the data for Switzerland of the form Y = a + b log X. The scatter diagram shows that this fits the data well. On the other hand Fuchs fits a regression of the form log Y = a + bX to the data for the U.S.A. Studies of U.S. data provide conflicting evidence. Fuchs fits a regression of the form Log Y = a + bX. If this form of regression were correct it would indicate that wage rates increase with city size at an increasing rate. On the other hand Mattila and Thompson (1968) fit a regression of the form Y = a + b log X which would indicate that wage rates increase at a decreasing rate.

In the preceding section we could treat land or floor space as a homogeneous good but it would not be correct to do this with labour. Different types of labour are paid at different wage rates, and it may be that the different wage rates will have different rates of increase with city size. There are some groundsfor expecting that although the absolute rate of increase of high wage rates will be greater than the absolute rate of increase of low wage rates, the proportional rate of increase will be lower.

We assume that we can use the theory of residential location to show that households of any given income group will always locate in the same relative position in any size city in which the characteristics of the households are the same i.e. if these households locate half way between the centre and the edge in one city they will in all cities. The compensation these households will require for living in a larger city will be equal to the increase in rent costs plus the increase in travel costs. For those living at the centre only the increase in rent costs is relevant and for those living at the edge only the increase in travel costs is relevant, and we would expect that the further the household lived from the centre the greater would be the importance of the increase in rents relative to the increase in travel costs.

There is evidence to justify the assumptions that, firstly, the value of time spent travelling is a constant fraction of the wage rate and this fraction does not vary with income, and, secondly, the average income elasticity of demand for housing is equal to one so that the amount spent on housing is a constant fraction of total income.[3] Therefore the amount necessary to compensate households in each income group for increased rents and increased time spent travelling will increase proportionately with income. On the other hand the amount necessary to compensate households for the increased direct financial costs of travel will not increase with income. Hence we would expect that the increase in wages necessary to compensate households for living in a larger city would be proportionately smaller but absolutely larger, the higher the household's income but furthermore because, as a rough approximation, the average income of resident households increases with distance from the centre in Anglo-American cities, it follows that the costs of increases in city size are mainly rent costs for poor households and mainly travel costs for rich households. Hence the total increase in costs with city size for poor households will tend to be nearly proportional to their income whilst the total increase in costs for rich households will be much less than proportional. Thus this argument tends to reinforce the previous argument.

There is little concrete evidence in support of the assertion that the rate of increase of high wage rates will be proportionately smaller but absolutely larger than the rate of increase of small lower rates. A casual study of the various 'London Living Allowances' which are incorporated in many British salary scales suggests that whilst they increase with income they certainly do not increase proportionately. Some evidence appears to conflict with the hypothesis. Farbman (1971) using data for S.M.S.A.'s in the United States found that family income inequality tends to increase with the size of the S.M.S.A. (in terms of population). This is rather less direct evidence however and its apparent contradiction of the hypothesis might be due to the fact that Farbman uses the Gini coefficient as a measure of inequality. Our hypothesis might be stated in statistical terms as a prediction that both the mean wage level and the standard deviation of wage rates about the mean will increase with city size, but the coefficient of variation will decrease. Since the Gini coefficient is not necessarily measuring inequality in the same way as the coefficient of variation, Farbman's findings do not refute the assertion.[4]

V Input Prices as a Function of City Size - Business Services

If the cost of labour and of floor space increases with city size whilst the cost of capital does not vary, there must be some compensatory reduction in costs with city size which allows the firms located in larger cities to remain profitable. In other economic discussions of the role of the large city authors have described various ways in which costs might be reduced in large cities. These economies of city size have been called 'external economies' (Lichtenberg (1960), Hoover and Vernon (1959), Vernon (1960) or 'urbanisation economies' and 'localisation economies' (Isard (1956)).

The most extensive empirical study of the nature of these 'external economies' was a product of the study of the economy of the New York Metropolitan Region under the direction of Raymond Vernon. From this study it would appear that these external economies are basically of two kinds.

In the first place there is the reduction in uncertainty which comes from location in a large city in close proximity to many possible sources of information. Vernon stresses that the manufacturing industries which cluster in the centre of the New York region are those which are characterised by extreme uncertainty.

> The shops and plants of these industries . . .are plagued by the fact that their output is unstable in one way or another. It may be unstable because of seasonality. But more important, as a rule, are the instability and uncertainty which come with a swiftly changing product or a highly variable demand (Vernon (1960) p.101)

Despite the possible economic importance of the reduction in uncertainty which may be possible in a large city we shall ignore this type of external economy here. In our analysis we shall concentrate on the second type of external economy. This economy results from the availability of a variety of specialised facilities and services in the large city.

> For example there are commercial laboratories which provide technical and research services on a contract basis; firms that design patterns for apparel manufacturers; agencies that maintain comprehensive files of pictures for magazines; and free-lance specialists in art work (Lichtenberg (1960) p.67)

Hence 'certain operations and services that a firm in a smaller place would have to do for itself can, in the city, be farmed out to separate interprises specialising in those functions and operating on a large enough scale to do them more cheaply'. (Hoover (1948)).

Now, as Isard (1956, p.182) remarks, the theory of urbanisation and localisation economies is in an unsatisfactory state. But we require a theory to explain the fall in the costs of certain services with city size (i.e. the external economies or urbanisation economies of city size). The approach we shall adopt is to regard urbanisation economies as a result of the greater division of labour possible in the large city. We can then bring to bear the theoretical analysis of the division of labour. This approach seems to be justified in that the descriptions of these economies by Lichtenberg or Hoover quoted above are exactly descriptions of the process of the division of labour. Moreover other students of the city have also equated the division of labour with the external economies of city size. Lampard (1969), for example, has commented on its importance as a cause of the growth of cities.

As Lampard observes, however, "division of labor . . .has not excited much theoretical interest or sustained observation. Division of labor, to be sure, was central to Adam Smith's scheme for enlarging the wealth of nations and he devoted all of three or four pages to the topic. Since Smith there have been no more than half-a-dozen noteworthy discussions of the topic". One of them is a paper by Stigler (1951) in which he clarifies Smith's theorem that 'the division of labour is limited by the extent of the market'. We shall use his method of analysis. We first present an outline (and extension) of Stigler's analysis, then use it to understand the nature of the 'external economies' of city size and to explain the way in which the cost of business services will vary with city size.

In Stigler's view the firm can be conceived as engaging in a series of distinct operations or activities or as being composed of a number of functions. Those familiar with the literature of urban economies will recognise this idea as being identical to that proposed by R. M. Haig in his classic 1926 paper, namely that the firm is viewed as consisting of a bundle of functions. The costs per unit of production of these activities may either be related or independent. Here we assume (as Stigler does) that the costs are independent, and therefore that the cost per unit of output of each activity depends only on the rate of output of that activity. It is also necessary to assume that the output of each activity is a constant proportion of the output of the final product by the firm. The purpose of making these assumptions is that they allow the cost curves of all the functions to be drawn on the one diagram and the conventional long run average cost curve of the firm to be drawn by summing vertically the cost curves of the various activities.

We should expect to find that the (long run) average cost curve of each of the functions of a firm would have a different shape. The three variations are shown in Figure 5, where the output of the firm's final product is indicated

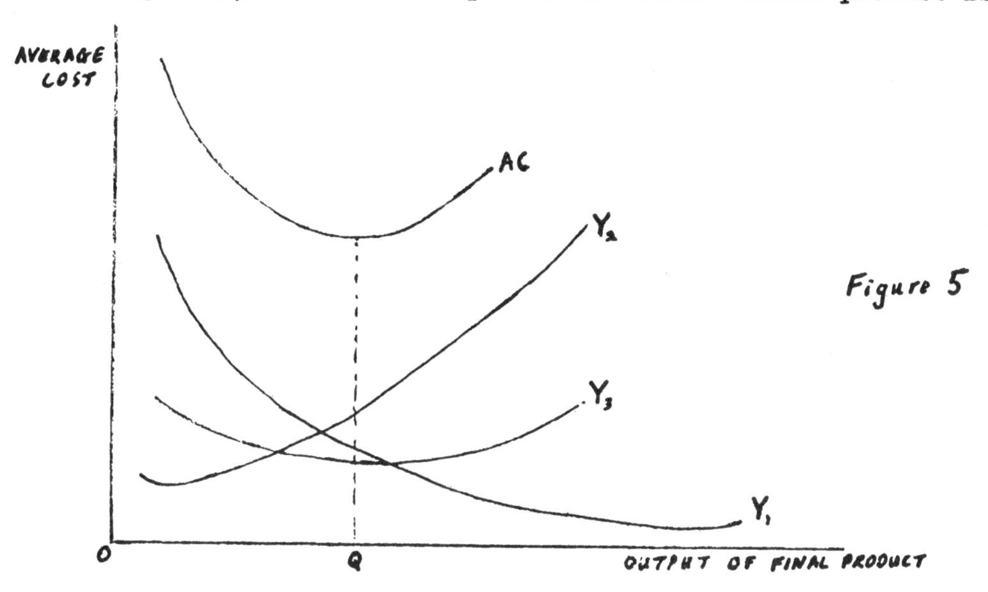

Figure 5

on the horizontal axis, and the average cost of the output of each activity or of the final product of the firm is indicated on the vertical axis. The cost curves of some functions fall throughout the probable range of the firm's output (Y_1); some rise continuously (Y_2); some have the conventional U- shape over that range, indeed determine that range (Y_3). The average cost of production of the final product of a firm which used these three activities is shown in Figure 5 as AC. The lowest point on this average cost curve occurs when the firm's output is Q. If all three activities have to be carried on by the same firm Q would be the firm's optimal output.

If all three activities don't have to be carried on together some readjustment is possible in order to reduce the cost of production. We assume that the long run average cost curves of the functions shown in Figure 5 will be the same whether these functions are carried on by this firm or any other and whether they are carried on singly or in multiples.[5]

The first thing that the firm can do is to produce the output of activity Y_2 from several processes and not from one single process which is subject to decreasing returns to scale.[6] If the activity or process Y_2 has a minimum cost at an output of Q_2 where $Q_2 < Q$, the firm may be able to use the activity in multiples in order to minimise its average costs of production. If the activity Y_2 can be carried out in multiples at constant (or nearly constant) cost, the average cost to the firm of the output of Y_2 and the average cost of the final output can be reduced.

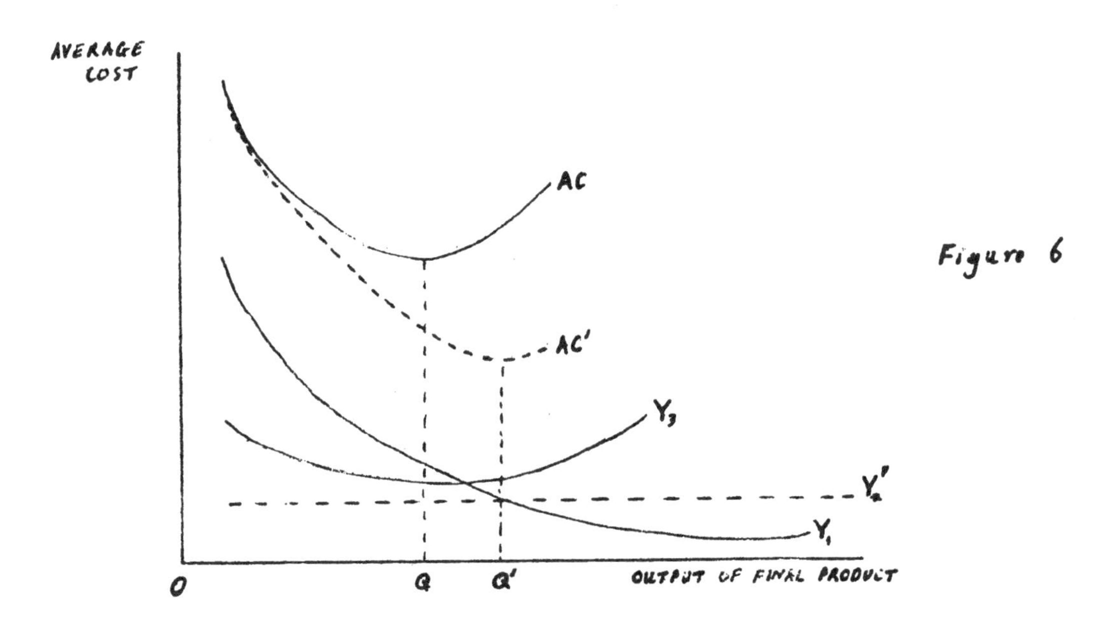

Figure 6

The new situation is shown in Figure 6. The curves marked AC, Y_1, and Y_3 are reproduced from Figure 5. The new constant cost curve for the activity Y_2 if it is carried out in multiples is marked Y_2^1. The new average cost curve for the firm is marked AC^1. It lies an increasing amount below the old average cost curve AC because the amount by which the average cost of the final product is reduced by using multiples of Y_2 is an amount which increases as the output of the firm increases. Thus the minimum average cost of the final product will be reduced and the optimal output Q^1 will be greater than Q since the lowest point of AC^1 will lie to the right of the lowest point of AC[7].

The second way in which the firm could reduce its costs is by varying the output of activity Y_1. It is obvious that it would be to the firm's advantage to allow processes like Y_1 to be run to produce an output greater than that necessary for the firm's production of its final product whether this output is Q or Q^1. This can only be done if the market for the production of the activity is greater than Q (or Q^1): that is to say, it can only be done if the excess production from the process can be sold to other manufacturers.

If this is possible there is no necessary reason for the process Y_1 to be controlled by any one of the firms using the output. Indeed since the firms buying the output of the process will usually be rival manufacturers, they will certainly prefer purchasing from an independent firm to purchasing from a plant owned by one of their own number. Therefore a new firm (firm A) will be set up using only the process Y_1 and this firm will sell the production from the process both to the firm represented here (firm A) and to other firms. Firm A could then purchase what it required of the output of process Y_1 from firm B at a price lower than the cost to firm A of running the process Y_1 itself. Hence the final product of Firm A can be produced at a lower average cost.

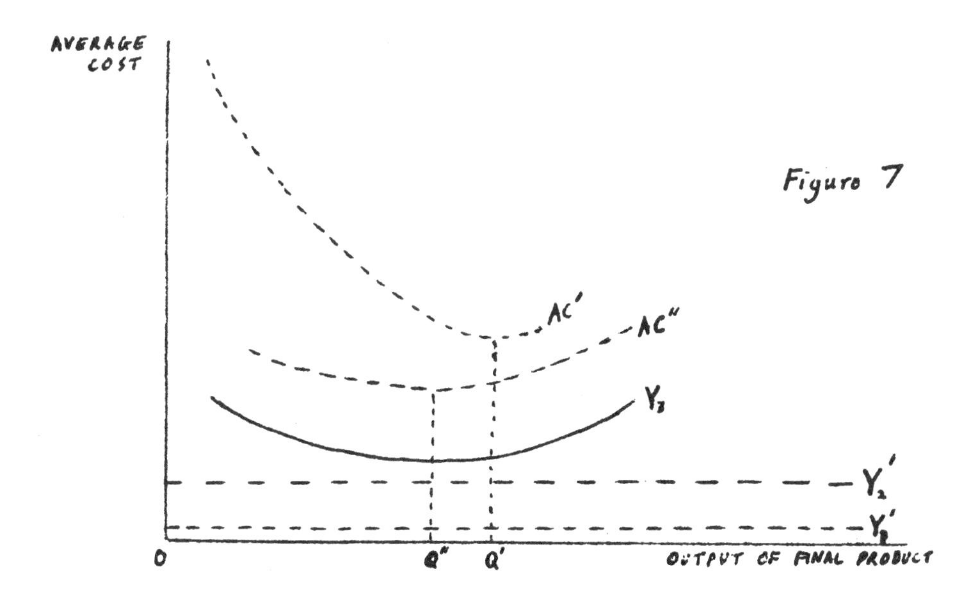

This result is illustrated in Figure 7. By buying in what the firm requires of the output of the process Y_1 the cost of the input from Y_1 is reduced to a constant level Y_1^1. The process Y_2 is still assumed to be run in multiples and the product of that process is therefore available at a constant cost Y_2^1.

The process Y_3 still has the same (long run) average cost curve. The average cost curve for the final product of the firm is lowered by the firm's ability to buy out but is lowered by an amount which decreases as the output of the firm increases. The average cost curve falls from AC^1 to AC^{11}; furthermore the new optimal output Q^{11} will be smaller than Q^1 since the lowest point of AC^{11} will lie to the left of the lowest point of AC^1. This follows necessarily from the fact that the amount by which the average cost is reduced by purchasing from outside is an amount which decreases as the firm's output increases.

It is clear that Stigler's analysis yields an explanation of the nature of certain of the external economies or urbanisation economies of cities. Those functions which have an optimal output greater than is required as an input by a single firm will operate as separate firms. This will be particularly true of services which are required relatively infrequently by most firms, e.g. the advice of accountants and lawyers. These firms will tend to locate where they can be sure that the demand will be great enough for them to achieve maximum economies of scale, i.e. in the centre of the city. Thus we would expect the range of business services to be greatest in the largest city in the economy. Partial confirmation of this result is provided by Clemente and Sturgis (1971) finding that there is statistically significant correlation between industrial diversification and city size.

The theory of intra-urban location lies outside the scope of this paper but we may note that if since firms providing specialised services will tend to locate at or near the centre of the city, the smaller firms which rely on these services will tend also to locate at or near the centre since they will be willing to pay the high rent costs associated with location near the centre in order to gain proximity to these services.

As the scale of a firm expands and more and more activities are incorporated within the firm's production system, so its ties with the firms providing business services at the centre become weaker and the transport costs of the firm to and from these firms also fall. As they fall the profit maximising location of the firm becomes more distant from the centre. Hence if the theoretical analysis is correct we would expect to find that the average size of firms would tend to increase with the distance they are located from the city centre. This of course is what Martin (1969) found in Greater London and Hoover and Vernon (1959) found in the New York Region:

> If we look at the 101 industry groups in the Region for which comparisons can be made in 1956, we find that in 83 of them the average plant in New York City was considerably smaller than the average plant outside the city.

Returning to the main theme of the argument, for the purpose of construction of the theory we wish to assume that the whole range of business services is available in all cities, but that the cost of any service will tend to decline with city size as the demand for the product increases and economies of scale become possible. What is assumed to be absent from the system is the possibility that firms in a small town can use the business services of a nearby city which is larger. This 'hinterland' effect may be important in practice but is obviously difficult to allow for in a general theory of city size since the spatial system of cities will vary from country to country and region to region.[8]

We shall show that for a representative business service the average cost of production will be decreasing if the total production demanded is less than the optimal output of a single plant, that at rates of output greater than this it will sometimes be increasing but will usually be decreasing and at high rates of output it will tend towards a constant cost of production equal to the minimum cost of production in a single plant.

We assume that a firm operates a process or activity which can provide a particular service at a minimum average cost per unit of output equal to C when the output of the firm is equal to Q. For simplicity we also assume that costs increase and decrease linearly with output and that the average cost per unit of output of the service is equal to 1-5 C if the output is equal to 0.5 Q or 1.5 Q. Hence the average cost of the service will be 2C if 2Q is produced and will approach 2C as the output of the service approaches O.. The shape of the average cost curve of the firm is shown in Figure 8 below.

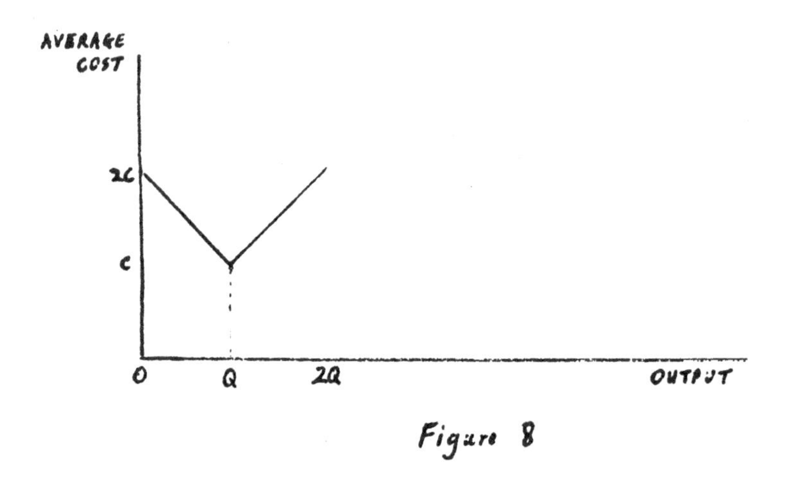

Figure 8

It is obvious that the output of 2Q can be produced more cheaply if two processes are used instead of one. When production is equal to NQ, and N is an integer, the average cost of the service will be C. It might be assumed from this that the average cost curve of the industry could be found by simply repeating the average cost curves of the firm as shown in Figure 9 where an output of (N + 0.5)Q is always produced at an average cost of 1.5C.

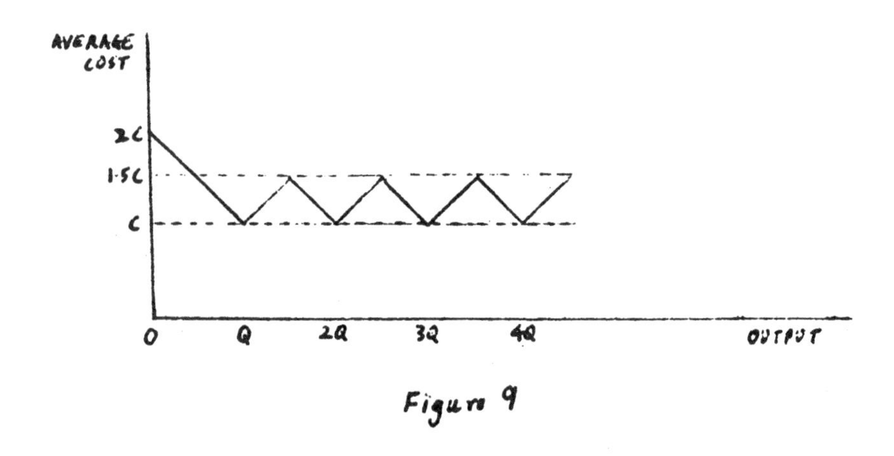

Figure 9

The average cost curve of the industry does not look like this, however. It is true that 1.5Q can be produced by one process operated on its own but it can be produced more cheaply if two processes are used, both producing 0.75Q. The average cost of production of 1.5Q is therefore 1.25C instead of 1.5C. Similarly the average cost of producing 4.5Q will not be 1.5C but 1.1C, whilst the average cost of producing 24.5Q will be 1.02C. The actual shape of the average cost curve of the industry is shown in Figure 10. As the number of

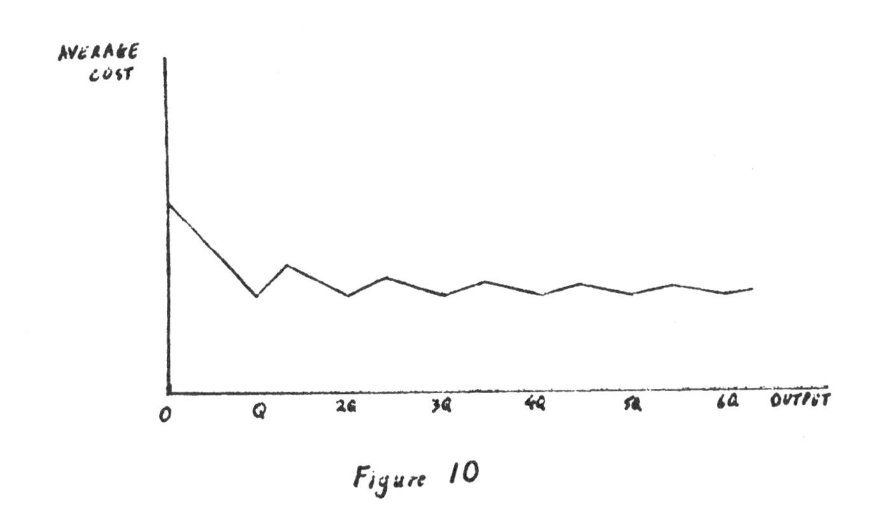

Figure 10

processes or plants being used to provide the service increases, the average cost curve of the industry tends toward a straight line and constant returns to scale, even though considerable economies of scale are available for the individual firm. When the average cost curve of the firm is of the conventional U-shape the average cost curve of the industry will tend towards linearity even more quickly, as shown in Figure 11, since only the flattened bottoms of the cost curves are relevant once the demand is great enough.[9]

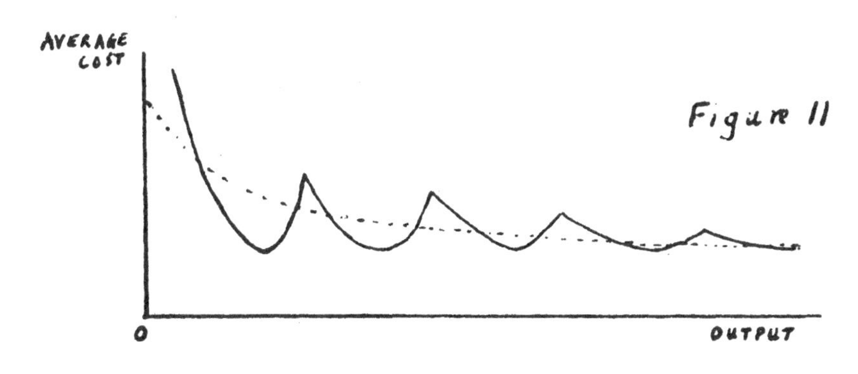

Figure 11

A curve of the shape shown in Figure 11 is difficult to handle analytically, but it seems reasonable to approximate the irregular curve by a smooth curve of the type shown as a dotted line in Figure 11. The trend curve has the property that it passes "through the average ordinate of each section of the [irregular] curve half-way between its consecutive minima. . . .The trend curve decreases continually at a decreasing rate and is asymptotic to the horizontal line of constant minimum cost" (Joseph (1933) p.394)

Since the demand for a business service will be positively correlated with city size, it follows that the trend curve shown in Figure 11 might also be taken to show the cost of the service as a function of city size. There is one qualification which must be made, however. Since land and labour will be used in the production of the service, the cost of the service as a function of city size will not decrease continually but will start to increase at some point, first at an increasing rate and then at a decreasing rate, as shown by the curve AA^1 in Figure 12.

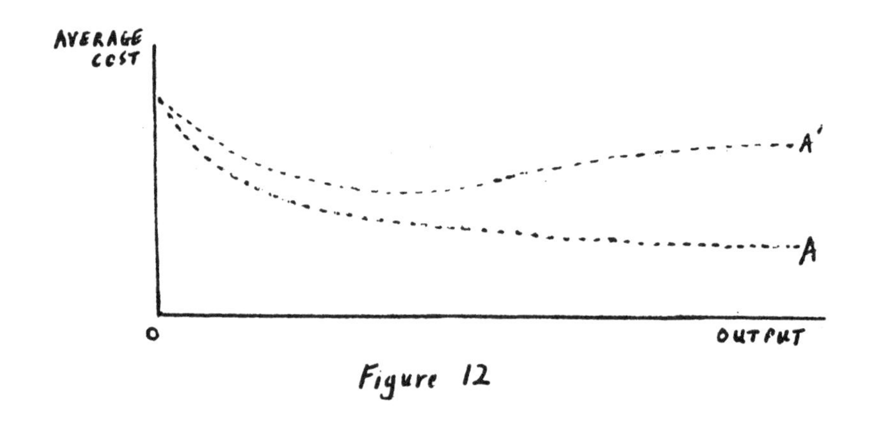

Figure 12

The curve AA shows the cost of supplying the service if the cost of land and labour did not vary with city size. When the curve AA is nearly horizontal the shape of the curve AA^1 is determined by the variation in the cost of land and labour with city size, and the curve slopes upward at a decreasing rate.

May 25th 1971

The cost of supplying each of the various business services as a function of city size will be different and can be represented by a different supply curve. Services which can achieve all the attainable economies of scale with a small output will be supplied at minimum cost in a small town or city, particularly if the production process uses little capital and a lot of land and labour. On the other hand services which require a large output to achieve all the economies of scale and which use a lot of capital and little land and labour will be supplied most cheaply in the largest city in the economy and even there all the economies of scale may not be achieved. Three possible cost of supply curves are shown in Figure 13.

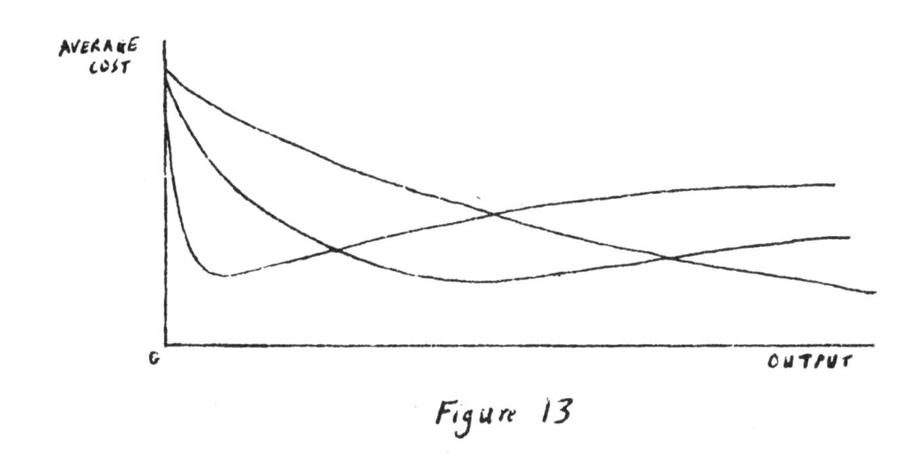

Figure 13

VI <u>The Locational Choice of the Manufacturing Firm - Partial Equilibrium</u>

For the purpose of construction of the theory we have divided the firms in the economy into two types; 'manufacturing' firms which operate at constant returns to scale, and for which the cost of transporting the final product is zero; and firms providing 'business services' whose plants are indivisible and which therefore operate at increasing returns to scale, the service being transportable only at an infinitely high cost.[10]

By assumption, all business services are available in all cities, though, as the analysis in the preceding section shows, the cost of business services will vary with size being highest in the smallest towns and cities and tending to decrease with city size. On the other hand we assume that each manufacturing firm can choose in which of the cities in the economy to locate. Since the cost of transport of the products of the manufacturing firms is zero and the sales of the firms do not vary with their location, it follows that the most profitable location will be the one at which their manufacturing costs are least.

Consider the locational choice of the individual firm. From its point of view the size distribution of cities in the economy is given, and the relative costs of location in each size of city can be ascertained. The firm will tend to locate in the size of city in which its costs are minimised and profits maximised. It will do this either because it deliberately chooses to locate there or because, if it does not, it will be driven out of business by firms which are located there. In the latter case the process of natural selection ensures that the location of the firm is the same as if it had deliberately set out to maximise profits. For the sake of brevity therefore we assume that the location of the firm is the result of a deliberate choice.

We also assume initially that the firm's inputs are used in fixed proportions. Thus a manufacturing firm uses specified quantities of floor space, labour, business services and capital to produce its output. The prices of these inputs vary with city size as shown in the two previous sections. By calculating the cost of each input as a function of city size and adding the various costs together the firm will be able to find the city size which minimises its costs.

The calculation is illustrated diagrammatically in Figure 14. The cost of the firm's floor space as a function of city size is shown in part (a), the cost of labour costs is shown in part (b), the cost of the business services it uses is shown in part (c), and the cost of capital is shown in part (d).

The curve showing its total costs as a function of city size is shown in part (e), and is found by summing vertically the curves in parts (a), (b), (c), and (d).

When cities are small, the total costs of the firm fall as city size increases because the decreasing cost of business services outweighs the increasing cost of labour and floor space. At some city size total costs are

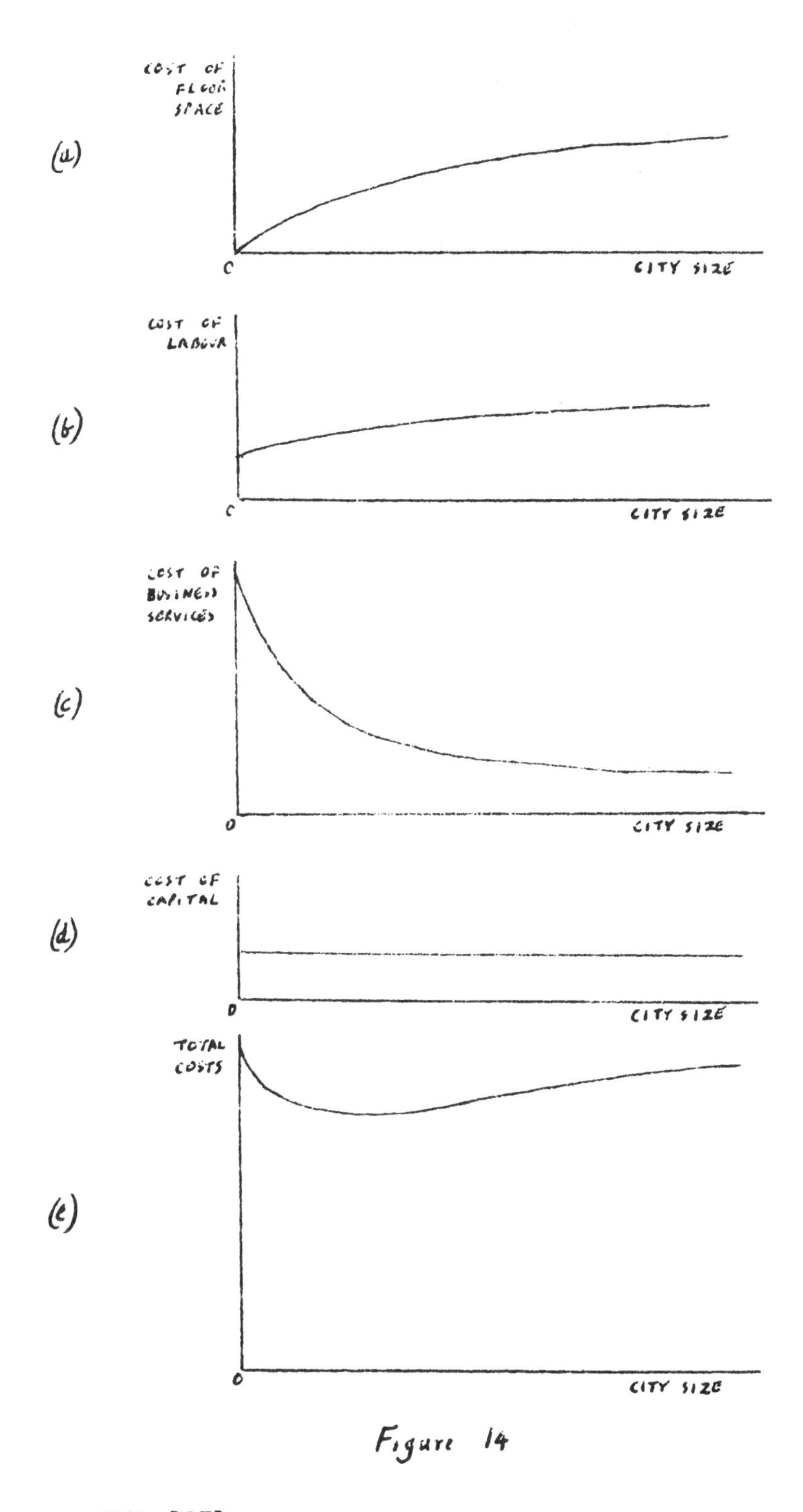

Figure 14

constant for small changes in city size and this is the firm's optimal location; the rate of decrease of the cost of business services exactly equals the rate of increase of the cost of labour and floor space. When cities are larger in size the increasing cost of land and labour outweighs the decreasing cost of business services and the firm's total costs increase as cities increase in size.

It may be noted that a stable equilibrium location at some intermediate city size is only possible if the cost of at least one of the three inputs, land, labour, or business services, either decrease at a decreasing rate or increase at an increasing rate. In sections III and IV we showed that the price of land and labour are expected to increase at a decreasing rate so that it was essential to the analysis to show in section V that the cost of business services should decrease at a decreasing rate. If, for example, the cost of business services decreased at an increasing rate, there would only be two possible city sizes, very small and very large, since all firms would find it unprofitable to locate in a city of any intermediate size.

For each manufacturing firm there is, however, an optimal city size, and location in a city of this size reduces the firm's costs to a minimum and maximises its profits. Obviously since in any national economy there are not cities of every possible size but only a limited number, most firms will not be able to locate in a city which is optimal in size for them. A firm which cannot locate in the optimal size city will have to locate either in the largest city which is smaller than the optimal size or in the smallest city which is larger than the optimal size. In one of these cities (and, possibly, but not necessarily, in both) its costs will be minimised and profits maximised. This city size we call 'nearly optimal'.

The optimal, or nearly optimal, city size will not be the same for all types of manufacturing firm but will depend upon the proportions in which they use the various kinds of input. At one extreme, firms or plants which use large quantities of land and low wage labour relative to their use of high wage labour, capital and business services will locate in the smallest cities. An example of this type of firm would seem to be firms in the textile industry which require cheap labour and tend to locate in small towns. Morse, Mathur, and Swamy (1968) show that the textile industry in the U.S.A. tends to be concentrated in small towns.[11] British Census data shows that this is also true in Great Britain.

Table 1

Proportions per 10,000 persons in Employment by Industry, England and Wales, 1961

Industry Order or Minimum List Heading	Six Conurbations	Areas outside conurbations			
		Urban areas with population of			Rural Districts
		100,000 or more	50,000 and less than 100,000	less than 50,000	
Order X Textiles	377	304	212	448	138
M.L.H.411 Production of man made fibres	5	42	18	36	16
412 Spinning and doubling of cotton, flax, and man-made fibres	9	81	15	28	58
413 Weaving of cotton, linen, and man-made fibres	30	35	47	113	30
414 Woollen and worsted	151	9	29	35	17
423 Textile finishing	39	26	8	42	12

Source: Census 1961, England and Wales, Industry Tables part I (London:H.M.S.O. 1966) Table A.

Table 1 shows that the proportion of the working population employed in the textile industry is greatest in urban areas with populations less than 50,000. It also shows that it is second greatest in the conurbations but this is misleading since the cotton towns of Lancashire and the woollen towns of the West Riding of Yorkshire are included in the South East Lancashire and West Yorkshire conurbations though for the purposes of the theory they should be considered as individual towns.

At the opposite extreme, firms or plants which use large quantities of business services, capital, and high-wage labour relative to their use of low wage labour and floor space will locate in the largest cities. Example of this type of firm might be found in the insurance, banking, and finance complexes which dominate the centres of the largest cities although it is difficult to know whether firms in the complex constitute 'business services' subject to indivisibilities or are the firms using the business services. A better example of plants of this type are the head offices of manufacturing companies which tend to concentrate in the largest cities of any national economy. Thus 52% of the five hundred largest British industrial companies had their headquarters in London in 1969 (Evans (1970)) whilst 33% of five hundred largest U.S. industrial companies had their headquarters in New York City in 1963 (Goodwin (1965)).

If we relax our initial assumption that the firm's inputs are used in fixed proportions, and allow the proportions to be varied, then we cannot use the diagrammatic analysis. A mathematical solution to the problem is possible, however, though little is added to the analysis. At the cost minimising or profit maximising location, once the firm has adjusted its inputs to take account of the price structure ruling in that city, any move to a slightly smaller or slightly larger city will result in no increase in costs. The increase in rent and wages and, possibly, in the cost of some business services will be just equal to the decrease in the cost of other business services. Any move to a distinctly larger or distinctly smaller city will result in an increase in costs, however.

May 25th 1971

One point of interest which does arise when firms can vary the proportions in which they use inputs to take account of differences in relative prices is that, as prices vary, firms will tend to substitute cheaper inputs for dearer. In the large city business services, capital, and high wage labour will tend to be substituted for floor space and low wage labour. Thus we would expect to find, for example, that the proportion of the working population in the higher paid occupations increases with city size. Mathur (1970) demonstrates that this is true for S.M.S.A.'s in the United States.

VII The Urban Hierarchy - General Equilibrium

In the previous situation we dealt with the locational choice of the individual firm in partial equilibrium. In this section we attempt to show how the locational choices of all the manufacturing firms in the economy will interact to determine the size distribution of cities in the economy in general equilibrium.

The distribution of city sizes has been studied extensively by geographers and others starting with the pioneering work of Singer (1936) and Zipf (1949). These two authors suggested that the size distribution of cities generally corresponds to a rank-size distribution or the similar Pareto distribution. Later researchers (e.g. Berry (1961), Stewart (1958), Rosing (1966)) have questioned the general applicability of the rank-size rule. In particular Berry (1961) analysed the city-size distributions in 38 countries and showed that whilst 13 of the countries had rank-size distributions the remainder had either primate distributions or distributions intermediate between primacy and rank-size. Primacy exists "when a stratum of small towns and cities is dominated by one or more very large cities and there are fewer cities of intermediate sizes than would be expected from the rank-size rule" (Berry (1961)).

One characteristic which is general to every city-size distribution is that the number of cities in any size-class tends to increase as the average size of cities in any size-class is reduced; e.g. there tend to be more cities with populations between 300,000 and 400,000 in any economy than there are cities with populations between 400,000 and 500,000. If we use the term 'dense' to denote the number of cities in any size class, then we can say that empirical evidence shows that the distribution of city sizes decreases in density as average city size increases. It is this pattern which we intend to try to explain.

What we shall do here is to develop an intuitive approach to a solution of the problem. We are not yet able to give a rigorous mathematical proof although it is clear in what way the problem should be approached. Each city can be viewed as a 'club' or 'coalition' of which manufacturing firms are members. The firms in the economy can be viewed as players in an n-person game in which each firm acting independently can join coalitions of other firms and usually cannot be barred. The pay-off to each firm is a function of the size of coalition of which he is a member and the pay-off functions of the firms are not all the same. Research on the theory of economic clubs may, in due course, help to provide a rigorous solution of the city size distribution in general equilibrium but to date is has usually been assumed that all the participants have identical tastes. See Buchanan (1965), Pauly (1968)(1970). Other research which may be relevant is being carried out into the economic importance of the 'core', the allocation of resources in the economy which cannot be blocked by any coalition, but here the concern is usually with the situation as the number of participants tends to infinity.[12] See Aumann (1964), Vind (1965), Hildenbrand (1968). It is, however, obvious that there is a finite number of cities and a rigorous solution of the problem of the equilibrium urban hierarchy can only result from considering an economy in which there is a finite number of firms, so that most of the work on the core of an economy is not relevant.[13]

We shall demonstrate with two simple models that the city size distribution is likely under most circumstances to have the property mentioned earlier. We assume that each firm uses its inputs in fixed proportions and that labour costs, land costs, and the cost of business services are each a function of city size.

Then each firm's costs and profits are a function of city size and a graph of its costs can be drawn as in Figure 14d. We also assume that the size of a city is a linear function of the number of manufacturing firms in the city. Furthermore we regard each city as a coalition formed by a group of manufacturing firms. Each firm will tend to join the coalition of the size which minimises its production costs; either it joins it deliberately or by a process of natural selection, only those firms which are members of the optimal size coalition or of the nearly optimal size coalition survive.

Imagine that the economy starts without any cities but that each firm knows the costs of location in each possible coalition size (but not the costs for other firms). The firms would tend to form coalitions, initially on a random basis, later more purposefully. At the first stage those firms which wished to be single-member coalitions would stay apart but those which wished to join larger coalitions would immediately join with others. Any coalition which was larger than the rest at the end of the first stage would become the optimal location of those firms desiring to locate in coalitions of that size or larger who would then move to it. Other firms would move out of it on the third stage as it became too large for them and would join smaller coalitions. The process of movement and adjustment would continue until a size distribution of coalitions was reached. In most circumstances this would approximate to the sort of size distribution of cities which is usually observed, in that the density of the distribution would decrease as with increases in average city size.

We illustrate with two simple models.

(1) Suppose that each of the firms in an economy would minimise its costs by location in a coalition of a different size, and that there are as many firms as there are possible sizes of coalitions, i.e. if there are M firms in the economy the possible sizes of coalition are 1, 2, 3, . . ., M. Thus there is one firm whose costs are at a minimum if it is on its own, one firm whose costs are minimised in a coalition of two, one whose costs are minimised in a coalition o

three, &c. With each integer there is associated a firm which finds its lowest costs in a coalition of firms equal to that integer in number. For simplicity we can use these integers to name or number the firms.

We also suppose that if each firm is not in its optimal size coalition but in a coalition which is one firm larger or smaller then its costs are increased by x%, and that, more generally, if it is in a coalition which is n firms larger or smaller than the optimum then its costs are increased by nx%. A reasonably stable hierarchy of coalitions under these conditions is shown in Table 2 where the coalitions are shown in inverse order of magnitude.

Table 2

The Hierarchy of Coalitions (Model 1)

	Characteristics of coalitions												
Coalition size	1	2		5		12		29		70		169	
Marginal firms in each coalition by the numbers of the firms' optimal coalition sizes	1	2	3	4	8	9	20	21	49	50	119	120	288
Costs of the marginal firms in each coalition as a proportion of their optimal costs	1.0	1.0	1.1	1.1	1.3	1.3	1.8	1.8	3.0	3.0	5.9	5.9	12.9

The reason why the hierarchy of coalitions is reasonably stable can be seen in Figure 15. The firms are numbered along the horizontal axis; the numbers also indicate the optimal size coalition of each firm. The vertical axis indicates the costs of each member of each coalition as a proportion of its costs in its optimal coalition. The graphs show the costs of each member of the smaller coalitions shown in Table 2. The relative stability of the hierarchy will be realised if the straight lines of the graph are extended upwards. It can easily be checked that given the existing size distribution of coalitions every firm's costs would be increased by a move to any other coalition. The marginal firms in each coalition are listed in the second row of Table 2; these are the firms that would lose least by a move to the next smallest or next largest coalition

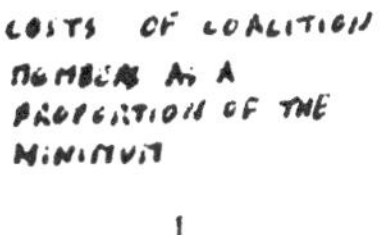

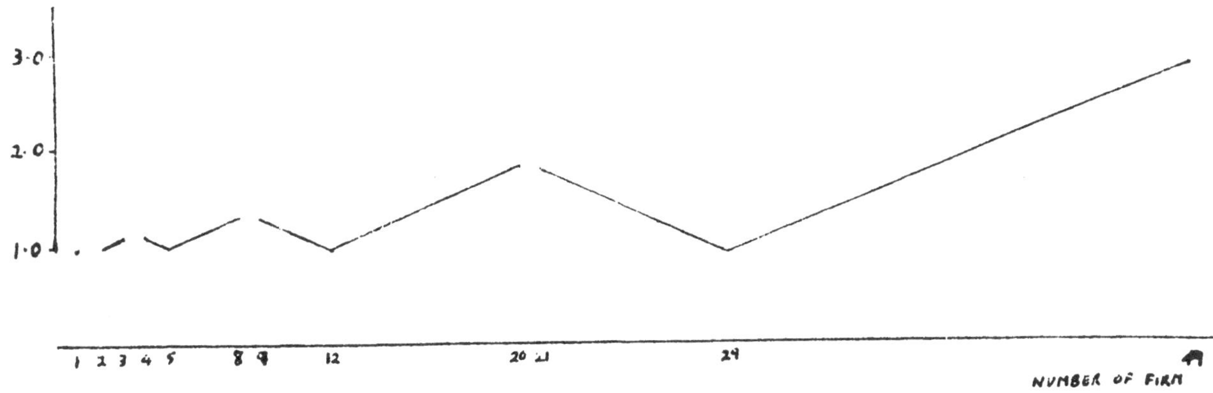

Figure 15

It may be noted that if firms are not short-sighted and realise that if they move to another coalition the size of that coalition will be changed then the hierarchy becomes rather less stable since marginal firms whose optimal coalition size is smaller than the one they are in will not be made worse off by a move to a smaller city so that the system may move away from equilibrium but should return to it.

Note that the hierarchy of coalitions is such that each one (above the third smallest) is equal to approximately 0.414 of the next highest,[15] so that the main feature of the usual city size distribution is reproduced, i.e. the density of the distribution decreases as the coalitions increase in size. In this model, however, the costs of the marginal firms diverge further and further from the optimum as the size of the coalitions increases. In the case of real world cities this is unlikely to be true, since, if it were true, variations in city size would be considered much more important by firms and the lack of cities which are intermediate in size between the larger cities in the economy would have been thought a serious problem. In the second model presented below this feature is absent.

(2) In the second model we again suppose that there is associated with each positive integer a single firm and that this firm would minimise its production costs in a coalition of firms equal in number to that integer. We now assume, however, that firms costs vary in a different way. We assume that if the firm's optimal coalition is N firms in size and it is in a coalition of N + n firms or N - n firms, then its costs will be $(N + n)/N$ times its minimum costs.

This pattern of cost variation seems more realistic than that in the first model. It states that equal proportional changes in coalition size result in equal proportional changes in costs, whereas in the first model equal <u>absolute</u> changes in coalition size result in equal proportional changes in costs. The earlier theoretical analysis of the rate of variation of input costs tends to confirm the new assumption. The rate of increase of land and labour costs as city size increases is much greater when cities are small than when they are large. Similarly the rate of decrease in the cost of business services as city size increases is much greater when cities are small. It follows that if a firm's optimal coalition size were small the proportional increase in a firm's costs resulting from its having to locate in a coalition a certain number larger or smaller than its optimum will be much greater when this optimum is small than when it is large. One can see intuitively that if a firm would minimise its costs by location in a city of 100,000, but it has to locate in a city of 1,100,000, its costs are likely to be proportionately higher than the minimum than if it wished to locate in a city of 10,000,000 and had to locate in one of 11,000,000.

May 25th 1971

Details of the hierarchy of coalitions which might be the result of firms attempting to minimise their costs under these conditions are shown in Table 3 and Figure 16.

Table 3

A Hierarchy of Coalitions (Model 2)

	Characteristics of Coalitions										
Coalition Size	1	3		9		27		81		243	
Marginal firms in each coalition by the numbers of the firms' optimal coalition sizes	1	2	4	5	13	14	40	41	121	122	364
Costs of the marginal firms in each coalition as a proportion of their optimal costs	1.0	1.3	1.3	1.4	1.4	1.5	1.5	1.5	1.5	1.5	1.5

Costs of coalition members as a proportion of the optimum

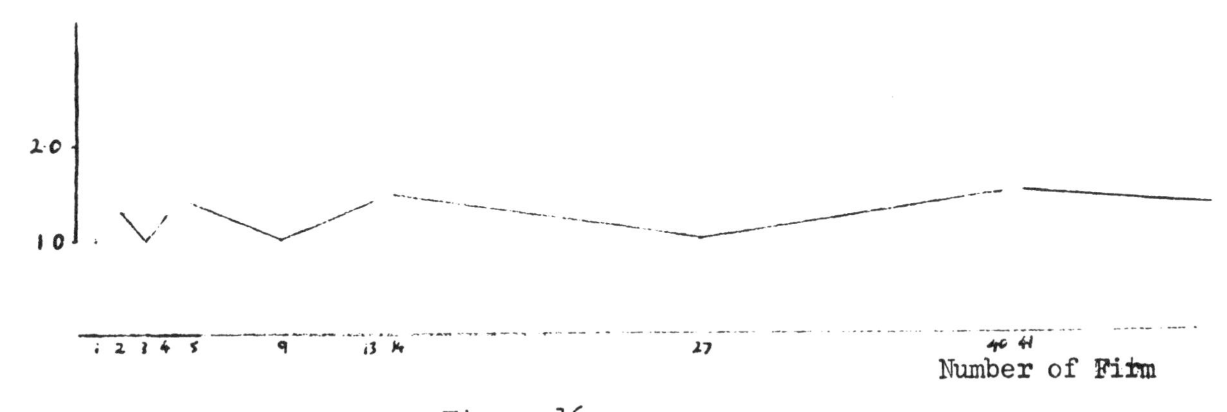

Figure 16

May 25th 1971

The most interesting feature of the second model as compared with the first is that the costs of the marginal firms do not diverge by increasing amounts from the optimum as the size of their coalition increases. Despite the existence of a hierarchy of coalitions in which each coalition is three times the size of the next largest, no firm's costs are greater than fifty per cent above the minimum.

It may be noted that if firms are not short-sighted and realise that if they move to another coalition the size of that coalition will be changed the hierarchy becomes rather unstable. It is always worthwhile for Firm 2 to join Firm 1 and then for Firm 1 to move again. A similar hierarchy could be formed of coalitions of 1, 2, 7, 21, 63, 189, &c. This has much the same characteristics as the hierarchy detailed in Table 3 but Firm 2 can no longer reduce its costs by moving. On the other hand Firm 4 can then reduce its costs by joining Firms 2 and 3 which puts Firm 2 back in its previous situation.

These stability problems are relatively unimportant, however. What we have shown with these two models is that an 'urban hierarchy' can result from the locational choices of individual firms each acting to maximise its own profits. Obviously a final conceptual leap is required to get from the notion of a hierarchy of coalitions to the idea of a hierarchy of cities, but, after allowing for variations in the size of firms and the existence of firms in business services and consumer services, the above analysis does provide an explanation for the existence of urban hierarchies in industrial economies. Furthermore this explanation appears to be relatively robust for, despite the differences in the assumptions made in the construction of the two models, a recognisable hierarchy is obtained in both.

There would appear to be only two cases in which we would not obtain a hierarchy of coalitions recognisably similar to the usual urban hierarchy. In the first place, if all the manufacturing firms required all the inputs in identical proportions, their cost functions would be identical and the result would be a number of coalitions (or cities) identical in size. In the second place the hierarchy of coalitions would not decrease in density as the coalitions increased in size only if the number of firms for whom N was the optimal size coalition was directly proportional to N. In that case the hierarchy would be uniformly dense. Both these conditions are highly unlikely so that a hierarchy of coalitions with the sort of characteristics usually observed in the urban hierarchy is the most likely form of hierarchy.

The actual hierarchy in any economy will depend upon the size of the economy (i.e. the number of firms in the economy) and the state of technology (i.e. the proportion of the firms in the economy for which each coalition size is optimal and the variation in their costs with city size). As an example a simple doubling of the number of firms in the economy represented in the second model, so that there are two firms associated with each positive integer instead of one results in a change in the hierarchy of coalitions from 1, 3, 9, 27, 81, &c., or 1, 2, 7, 21, 63, &c., to 1, 1, 2, 4, 6, 10, 16, 28, 46, &c., which is a considerably denser hierarchy than the others. Thus the number of coalitions with less than 30 members increases from four to eight if the number of firms in the economy is doubled. Other results of the increase in the size of the economy are that the largest coalition (or city) increases in size and the maximum divergence of any firm's costs from the minimum is reduced, in the example from 50% to 25%.

By manipulating the number of firms in the economy and the proportions finding their optimal location in each coalition size it would no doubt be possible to generate a hierarchy of coalitions which obeyed the rank-size rule. Possibly this could be done using simulation techniques. Another line of research might be to attempt to find the size-distribution which is maximally non-committal with respect to the missing information on the number and cost characteristics of firms, and this might be done using the methods of entropy maximisation. See Wilson (1970).

May 25th 1971

One way in which the discussion in this section might be modified and improved would be to relax the assumption that the inputs of each firm are fixed in proportion. We should then allow cheaper inputs to be substituted for dearer as the price structure changes with city size. It is probable that a hierarchy of the expected type would still result, however, whilst the complexity of the problem would be increased enormously and an analytical solution of this sort of n-person game would be extremely difficult.

VIII Normative Aspects of the City Size Distribution

Even when we can show that an urban hierarchy will come about if each firm in the economy attempts to maximise profits, we still would like to know whether this hierarchy is in some sense socially optimal. It can be shown that in a world-of-equals economy in which each firm wishes to locate in the same size of coalition then the economy will be Pareto optimal provided that the number of firms in the economy is divisible without remainder by the number of firms in the optimal size coalition (Pauly (1970)). This is a highly artificial and unrealistic case, however, and whilst Pauly conjectures that the 'tendencies' described by the world of equals model may also hold for non-homogeneous cases, considerably more work is required before we can completely understand the welfare economics of the urban hierarchy.

If we allow firms to adjust their inputs to minimise the effect of variations in relative prices as city size increases then it can be shown that the competitive allocation of resources will definitely not be Pareto optimal, since business services will be under-utilised and land and labour will be over-utilised in any city.

We consider business services first. Any expansion in the demand for business services will allow the firms providing the services to reduce the average cost of the services as they take advantage of the possible economies of scale. If a firm in any city expands its output then its demand for business services will increase and this will result in a reduction in the cost of business services. This reduction in the cost of business services will benefit both the firm expanding its output and all the other manufacturing firms in the city. Thus an expansion in the firm's demand for business services causes external economies, but these external economies are not taken into account by the firm when it decides its optimal output and its use of business services. Because of this the use of business services in any city will be less than would be socially optimal.

Contrast this with the use of labour. Any expansion in the use of labour will lead to an increase in wage rates since the increase in the population will lead to an increase in the area of the city and hence to an increase in rents and transport costs. Wage rates will therefore increase to compensate for increased living costs. The increase in wage rates and rents will raise the costs of both the expanding firm and all the other manufacturing firms in the city.

The same argument applies to increases in the use of floor space. Any expansion in the use of floor space by firms takes it away from residential use and causes a compensating increase in the area of the city. As a result rent, travel costs and wage rates increase.

To the extent that these increases in rents and wage rates reflect increases in the value of land (land, itself, not floorspace) they only result in a transfer of capital value and not in a diversion of resources. These capital transfers are merely pecuniary external diseconomies and do not cause the economy to diverge from Pareto optimality. On the other hand, to the extent that they reflect increases in travel costs and building costs they are technological external diseconomies causing a diversion of resources which results in the economy diverging from optimality. Because of this the use of labour and floor space in any city will be greater than would be socially optimal.

Two points of interest arise out of this discussion. In the first place there is no suggestion that there is some single optimal size of city. Instead the analysis implies that there is some optimal city size distribution which may or may not differ from the existing hierarchy. In the second place it is argued above that business services are underused and labour and floor space is overused in all cities both large and small. Indeed, since the cost of business services declines with output <u>at a decreasing rate</u> and wages and rents increase with city size <u>at a decreasing rate</u>, it may be that the divergence from optimality is greater in small cities than in large cities. This conclusion runs counter to much theorising on city size in which it is usually argued that very large cities should be reduced in size implying that the deviation from optimality increases with city size (above a certain size). The difference in conclusions may of course result from the fact that previous researchers have usually been concerned about congestion costs whilst we assume that the pricing problems of the transport system are solved and that there is no unpriced congestion. See Neutze (1965), Reynolds (1966).

There is, however, one way in which, using the theory of city size, we may show both that large cities may become too large and that new towns ought to be built. Suppose that in the course of an economy's development new firms come into existence which would minimise their location costs in very small towns. These firms could move into existing small towns but if they did it is probable that, initially, they would have to pay very high wages because the supply of labour would be inelastic in the short run. They would have to draw workers from surrounding towns and cities by high wages and also encourage workers to move to the small town. It may be that it would take a very long time for equilibrium to be reached during which time the cost of location in the town would be high. The firm could avoid these costs by location on the periphery of an existing large city. The firm would have to pay a wage which would draw workers away from the city centre, but the wage rate would not have to be as high as it would be in the short run in the small town whilst rents would be cheaper.

May 25th 1971

Furthermore some adaptation to the changed situation would take place within the city. Workers would change their residences to live nearer their work and would move to new housing built on the periphery beyond the new industrial firms. The pattern of location before and after the process of relocation is shown in Figure 17. In equilibrium the firm's rent would have increased and its wage rates would have decreased until they were both slightly above the equilibrium rate in a small town. If other firms locate in the same area the cost of business services will tend to fall until it too is slightly above the cost in the small town in equilibrium.

Figure 17

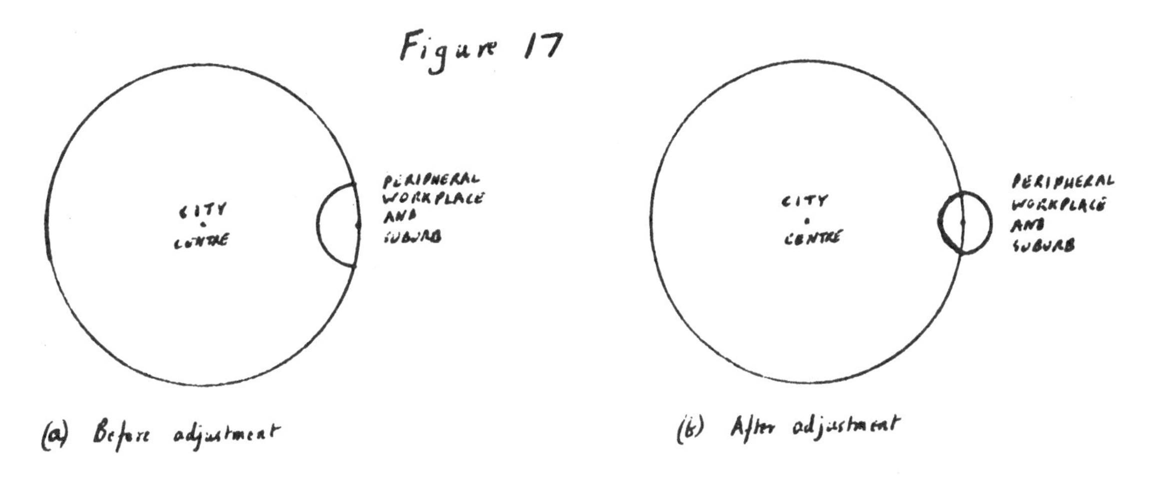

(a) Before adjustment (b) After adjustment

The increase in the cost to the firms at the centre of the large city will not be too great unless the number of firms at the periphery is large and the city becomes ringed with industrial suburbs as in Figure 18. Because the drain on the labour supply is now very great, firms at the centre will have to pay considerably higher wages. Furthermore any increase in their demand for labour will result in a higher increase in wages and rents at the centre then would otherwise be necessary since the increased demand can only be met by the new workers living beyond and travelling to work across the new industrial suburbs. Obviously one way of solving this problem would be to construct new small towns and encourage or direct both the suburban firms and their labour force to locate in these new towns. The analysis therefore provide some justification for a 'new towns' policy and gives a possible answer to Alonso's question 'What are new towns for?' (Alonso (1970)). It must be borne in mind, however, that we

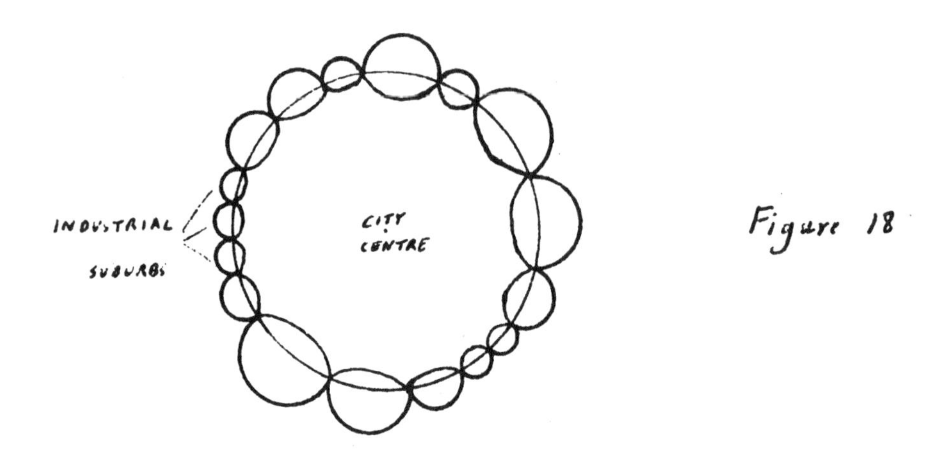

Figure 18

have no theory of the intra-urban location of firms and we therefore have not taken into account the benefits of location near to or within a large city but not at its centre. These benefits are not negligible. Therefore the above analysis does not provide an unqualified justification for a policy of 'overspilling' firms and people to new or expanded towns.

IX The Effect of Changing Transport Technology

What will be the effect of changes in transport speeds on the city size distribution? We shall show that if the speed of transport of commuters increases or the relative cost of commuting falls, then small towns will, on average, lose population whilst large towns will gain population.

An increase in the speed of transport of commuters or of a reduction in the relative cost of commuting will result in a fall in the slope of the rent gradient (Evans (1970) Chapter 11). Furthermore if the population of the city stays the same, the rent at the city centre and the wage premium required by workers to live in that size city will both fall. If the transport improvements occur in all cities simultaneously, these cost reductions will occur in all cities, though, obviously, the larger the city, the larger the reduction. Furthermore, because of the fall in rents and wage rates the cost of business services will also fall in the cities of every size, though, again, the fall will be larger in larger cities. The resulting change in the production costs of a representative firm is shown in figure 19.

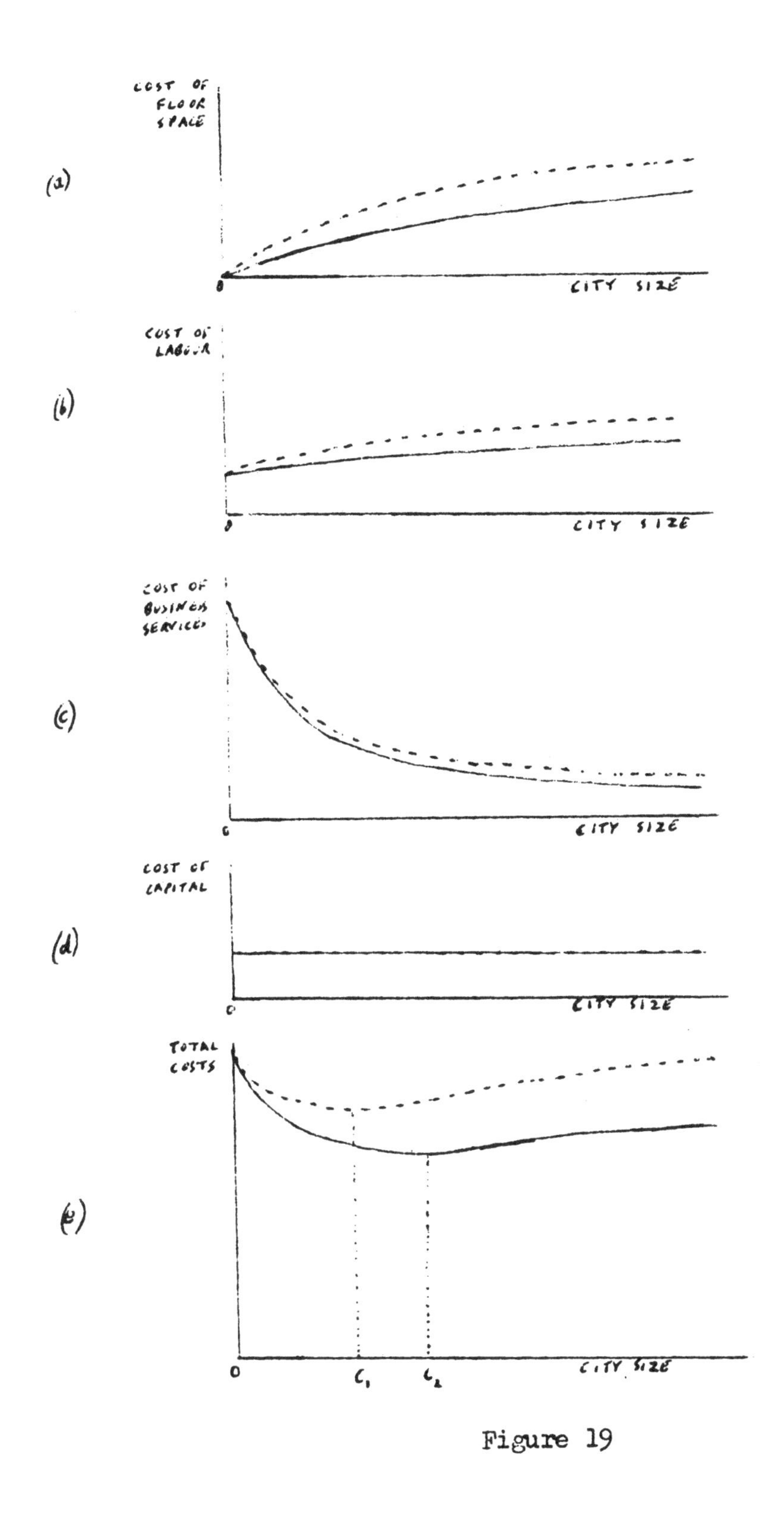

Figure 19

The cost of the firm's floor space as a function of city size is shown in part (a), the cost of labour is shown in part (b), the cost of the business services it uses is shown in part (c). The curves indicating the functional relationships before and after the change in transport technology are shown by dotted and solid lines respectively. The unchanged cost of capital is shown in part (d). The curves indicating the firm's total costs as a function of city size before and after the change are shown in part (e) and are found by summing vertically the curves in parts (a), (b), (c) and (d).

The total production costs of the firm fall at each possible city size but they tend to fall by an amount which increases with city size. Therefore the optimal size city for this firm will be greater after the change in transport technology than it was before it. In the Figure 19(c) this is shown by the fact that the minimum point of the solid line, C_2, lies further to the right than the minimum point of the dotted line, C_1. This will be true for every manufacturing firm; each one will find its minimum cost location in a larger city. How much larger will depend on the characteristic inputs of each firm, but it is clear that the smaller towns will generally lose firms (and population) whilst the largest will generally gain firms (and population). The proportionate increase or decrease in the size of cities as a function of city size predicted by this analysis is shown by the line AA in Figure 20.

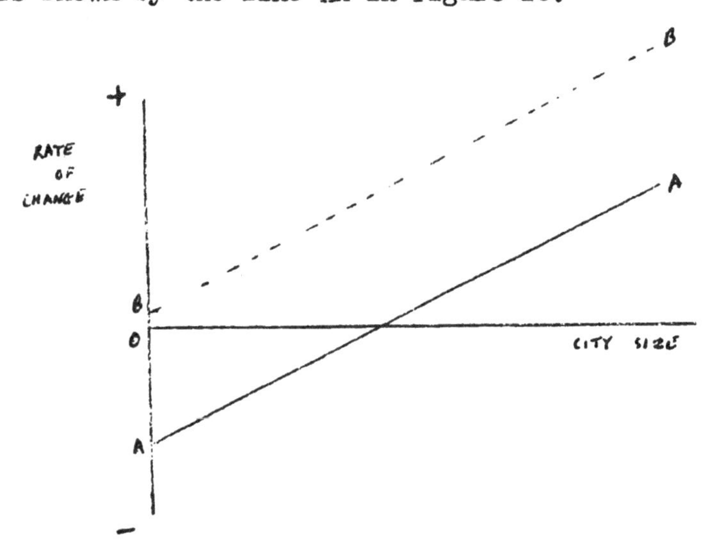

Figure 20

If the total population and the total number of firms is increasing over the period of adjustment of the economy to the change in transport technology, small towns are less likely to decrease in size and large cities are likely to increase still more in size. The changed relationship between growth and city size is shown in Figure 20 by the line BB.

It is interesting to compare this prediction with the rates of growth of U. S. S.M.S.A.'s in the decade 1950-1960. Stanback and Knight (1970) found that the rate of growth was greatest for medium-size S.M.S.A.'s rather than large size S.M.S.A.'s as shown in the graph in Figure 21. Furthermore the pattern of growth was the same in all the regions of the United States. It is probable that this result is due to the fact that the major transport improvements of the period did not affect all cities equally. The major change would be the increased use of automobiles and the beginnings of the construction of urban freeways. This type of transport improvement would affect large cities least since they usually have efficient public transport systems, congestion restricts the use of the automobile, and land costs restrict the construction of freeways. As a

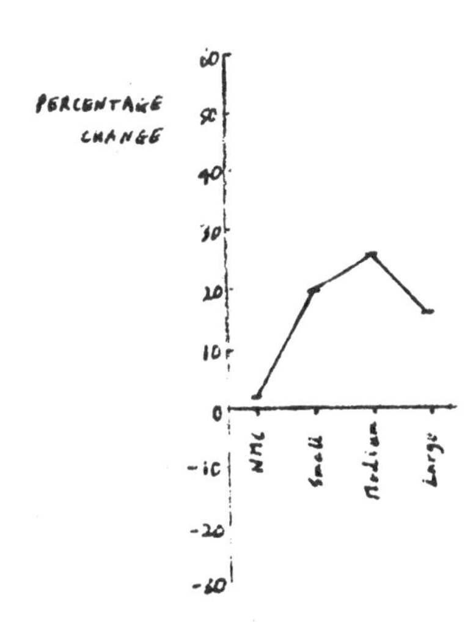

Figure 21

Source: Stanback and Knight (1970) p.97

Notes : NMC = Non-metropolitan counties
Small = S.M.S.A.'s with 50,001-200,000 population non-S.M.S.A. counties with a city of 25,000 or more
Medium = S.M.S.A.'s with 200,001-1,600,000 population
Large = S.M.S.A.'s with 1,600,001 or more population

result we would not expect the rate of growth of large cities to be as great as the rate of growth of medium-size cities, since in medium-size cities the transport improvements would have their greatest effect.

X In Conclusion

In this paper we have constructed a theory which explains the existence of urban hierarchies in industrial economies. In so doing we deliberately excluded the spatial aspects of the economy from consideration. We did this to simplify the construction of the theory and also to reveal the articulation of the theory more clearly. What we did not wish to imply is that city size is the only factor which enters into the choice of location of the manufacturing firm.

Other theories of location emphasise other factors, however; Weberian location theory tends to emphasize proximity to markets or to sources of supply; central place theory emphasises the importance of choosing the most profitable location to serve the market area or the supply area.

In the understanding of patterns of location in the real world all types of theoretical explanation must be taken into account. Thus a manufacturing firm will be located either near its raw materials or at its market; if it is located at its market, it will tend to locate near the centre of its market area in order to maximise its profits, but in choosing its location it must also weigh up the relative costs of location in each of the cities in the area in order to decide whether it is worth trading centrality for location in a larger or smaller city. Yet since each type of theory explains location in a different way and stresses some economic forces at the expense of others, each will be able to explain the location patterns of different types of plant, firm or industry.

In two major respects the theory presented in this paper is inadequate. In the first place it fails to take uncertainty into account. This may be an important omission. Lichtenberg (1960) and Vernon (1960) both stress the importance of location in New York City as a means of reducing uncertainty both through the great speed with which information can be obtained and through the fact that almost any business service is available on demand when it is needed. Uncertainty is a difficult concept to handle, however; developing a theory of city size which allowed for risk and uncertainty might be a difficult task. A possible line of approach might be to use the methods of portfolio analysis regarding each city as a 'portfolio' of firms.[17]

In the second place the theory fails to incorporate a theory of the intra-urban location of firms. The assumption that all employment is located at the centre of the city is a considerable simplification but it obviously conflicts with the facts. On the other hand it is true that the density of employment is highest at the city centre and declines steeply with distance from the centre so that the conflict is hopefully not too important.

Acknowledgements

I am indebted to Resources for the Future Inc. for financial support and to G. C. Cameron, L. C. Hunter, W. F. Lever, and Lowdon Wingo Jr. for their helpful discussion. In no case is any measure of agreement implied.

Notes

1. In this paper I "assume hierarchy to mean simply the order of size-rank", since I am concerned with the rank-size problem. This usage contrasts with that in "articles concerned with the functional level of cities which assume hierarchy to mean that restricted functions are higher in level (i.e. the-higher-the-fewer)"(Lukermann (1966)) So Parr (1970) would take the view that the theory advanced in this paper is nonhierarchical.

2. But it should be noted that Tinbergen (1968) has sketched out a type of central place theory which deals with manufacturing industry.

3. Thus British evidence suggests that the value of travelling time is equal to approximately 25% of the wage rate (Ministry of Transport (1969)). American evidence suggests that the income elasticity of demand for housing space must be approximately equal to one since it must be greater than 0.5, the income elasticity of demand for rooms, and less than 1.5 to 2.0, the income elasticity of demand for value of housing (Reid (1962)).

4. Other explanations of Farbman's result are also possible. It might be due to systematic variations in the pattern of residential location of income groups as cities increase in size coupled with a tendency for the area defined as the S. M. S. A. to exclude the fringe areas, or "exurbs" of the central city (Farbman (1971)). Alternatively it might be due to the substitution of higher paid workers for lower paid workers in larger cities as suggested in a later section of this paper.

5. I am indebted to L. C. Hunter for pointing out that this assumption is necessary to the analysis.

25th May 1971

6. It may be noted that Stigler does not deal with this case though it would appear to be implied by his analysis.

7. The implication of this result is that mass production leads to an increase in the scale of operation of the firm.

8. For a theoretical analysis of the effect of proximity to a large city on a smaller town's growth prospects see von Böventer (1970) and for an empirical study see von Böventer (1969).

9. The Figure is a reproduction of Figure IV in an article by M. F. W. Joseph (1933). See also Patinkin (1947) Figure 3.

10. The locational behaviour of the two other possible types of firms is relatively uninteresting and it would add nothing to the analysis to consider them. Firstly, there are firms for which there are economies of scale but which can transport their product at zero cost. The analysis of their locational pattern would not differ from that of the 'manufacturing' firm. Secondly, there are firms which operate at constant returns to scale but for which the cost of transporting the product is 'infinitely' high. These firms could be considered as providing business services.

11. But they note that this is "not evident in India except perhaps in incipient form".

12. As the number of participants increases the core shrinks to a single point which is both a competitive allocation and Pareto optimal.

13. But see Pauly's 1970 paper on 'Clubs and Cores'. It is worth noting that Pauly remarks that cities, among other things, could be profitably treated as clubs for the purposes of economic theory.

May 25th 1971

14. Thus suppose Firm 9 accidentally or deliberately moves to Coalition 5 making it 6. F(irm) 4 then gains by a move to C(oalition) 2 making it 3. F2 moves to C1 making it C2. Then F9 gains by returning to C11 making it C12, F4 returns to C4 making it C5, F1 leaves C2 and F2 rejoins F3.

15. If X_n is the number of firms in the nth smallest coalition, $X_{n+1} = 2X_n + X_{n-1}$ and $X_{n+1} = (\sqrt{2} - 1) X_n$

16. In a recent paper Lave (1970) correctly describes increases in travel costs as technological diseconomies but incorrectly describes all increases in rent costs as pecuniary diseconomies. Increases in the rent of floor space result in the diversion of resources into the construction of more floor space on each site.

17. The clearest exposition of portfolio analysis is given by Sharpe (1970).

References

William Alonso (1970), 'What are new towns for?' Urban Studies, 7 (February).

" " (1964), Location and Land Use (Cambridge, Mass; Harvard U.P.)

R. J. Aumann (1964), 'Markets with a continuum of traders', Econometrica, 32 (January)

B. J. L. Berry (1961), 'City Size Distributions and Economic Development', Economic Development and Cultural Change, 9 (July), partially reprinted in Berry and Horton (1970)

" " (1964) 'Cities as Systems within Systems of Cities', Papers and Proceedings of the Regional Science Association, 13.

" " (1967) Geography of Market Areas and Retail Distribution, (Englewood Cliffs, N.J.; Prentice-Hall, Inc.)

" " and F. E. Horton (1970), Geographical Perspectives on Urban Systems, (Englewood Cliffs, N.J.; Prentice-Hall, Inc.)

" " and Allan Pred (1961), Central Place Studies : A Bibliography of Theory and Applications, (Philadelphia: Regional Science Research Institute).

" " and M. J. Woldenberg (1967), 'Rivers and Central Places: Analogous Systems?' Journal of Regional Science, 7 (Winter).

Edwin von Böventer (1969), 'Determinants of Migration into West German Cities, 1956-61, 1961-66', Papers and Proceedings of the Regional Science Association, 23.

" " " (1970), 'Optimal Spatial Structure and Regional Development', Kyklos, 23 (4)

William Breit and Harold M. Hochman (1968), Readings in Microeconomics, (New York: Holt, Rinehart, and Winston, Inc.)

James M. Buchanan (1965), 'An Economic Theory of Clubs', Economica, 32 (February)

Frank Clements and Richard B. Sturgis (1971), 'Population Size and Industrial Diversification', Urban Studies, 8 (February)

Alan W. Evans (1970), Residential Location in Cities, unpublished manuscript.

" " " (1970), The Location of the Headquarters of Industrial Companies, unpublished manuscript.

Michael Farbman (1971), Family Income Concentration in Urban Areas in the United States, unpublished manscript.

Ronald Findlay (1970), Trade and Specialization (London: Penguin Education)

Victor R. Fuchs (1967), Differentials in Hourly Earnings by Region and City Size, 1959, N.B.E.R. Occasional Paper 101, (New York: National Bureau of Economic Research)

William Goodwin (1965), 'The Management Center in the United States', The Geographical Review, 55 (January)

R. M. Haig (1926), 'Toward an Understanding of the Metropolis: Part II', Quarterly Journal of Economics, 40 (May)

Robert Higgs (1970), 'Central Place Theory and Regional Urban Hierarchies: an empirical note', Journal of Regional Science, 10 (August)

W. Hildenbrand (1968), 'The Core of an Economy with a Measure Space of Economic Agents', Review of Economic Studies, 35 (October)

Edgar M. Hoover (1948), The Location of Economic Activity, (New York: McGraw-Hill

" " " and Raymond Vernon (1959), Anatomy of a Metropolis, (Cambridge, Mass.; Harvard University Press).

Walter Isard (1956), Location and Space Economy, (Cambridge, Mass.: The M.I.T. Press)

M. F. W. Joseph (1933), 'A Discontinuous Cost Curve and Increasing Returns', The Economic Journal, 43 (September)

Eric E. Lampard (1968), 'The Evolving System of Cities in the United States', in Issues in Urban Economics Harvey S. Perloff and Lowdon Wingo Jr., (Baltimore: The Johns Hopkins Press for Resources for the Future, Inc.)

May 25th 1971

lester B. Lave (1970), 'Congestion and Urban Location', Papers and Proceedings of the Regional Science Association, 25.

Robert M. Lichtenberg (1960), One-tenth of a Nation, (Cambridge, Mass.: Harvard University Press).

Fred Lukermann (1966), 'Empirical Expressions of Nodality and Hierarchy in a Circulation Manifold', East Lakes Geographer, 2 (August) Partially reprinted in Berry and Horton (1970)

J. E. Martin (1969), 'Size of Plant and Location of Industry in Greater London', Tijdschrift voor Economische en Sociale Geografie, 40 (November/December)

Vijay K. Mathur (1970), 'Occupational Composition and its Determinants: an intercity size class analysis', Journal of Regional Science, 10 (April)

John M. Mattila and Wibur R. Thompson (1968), 'Appendix: Toward an Econometric Model of Urban Economic Development', in Issues in Urban Economics ed. Harvey S. Perloff and Lowdon Wingo Jr., (Baltimore: The Johns Hopkins Press for Resources for the Future, Inc.).

Ministry of Transport (1969) 'The Value of Time Savings in Transport Investment Apparaisal', a paper prepared by the Economic Planning Directorate.

R. M. Morse, O. P. Mathur and M. C. K. Swany (1968), Costs of Urban Infrastructure as Related to City Size in Developing Countries (Palo Alto, Calif: Stanford Research Institute), partially reprinted in Berry and Horton (1970)

R. F. Muth (1968), 'Differential Growth among Large U.S. Cities', in Papers in Quantitative Economics ed. James P. Quirk and Avid M. Zerley (Laurence, Kansas: The University Press of Kansas).

May 25th 1971

R. F. Muth (1969), Cities and Housing, (Chicago: Chicago University Press

G. M. Neutze (1965), Economic Policy and the Size of Cities (Canberra: The Australian National University)

John B. Parr (1970), 'Models of City Size in an Urban System', Papers and Proceedings of the Regional Science Association, 25.

Don Patinkin (1947), 'Multiple-Plant Firms, Cartels and Imperfect Competition', The Quarterly Journal of Economics, 61 (February)

Mark V. Pauly (1968), 'Clubs, Commonality, and the Core', Economica, 35 (August)

" " " (1970), 'Cores and Clubs', Public Choice, 9 (Fall)

Margaret G. Reid (1962), Housing and Income (Chicago: University of Chicago Press).

D. J. Reynolds (1966), Economics, Town Planning, and Traffic (London: The Institute of Economic Affairs).

K. E. Rosing (1966), 'A Rejection of the Zipf Model (Rank-Size Rule) in Relation to City Size', The Professional Geographer, 18 (March)

William F. Sharpe (1970), Portfolio Theory and Capital Markets, (New York: McGraw Hill).

H. A. Simon (1955), 'On a Class of Skew Distribution Functions' Biometrika, 42 December

H. W. Singer (1936), The 'Courbe des Populations'. A Parallel to Pareto's Law', The Economic Journal, 46 June

Thomas M. Stanback and Richard V. Knight (1970), The Metropolitan Economy (New York: Columbia University Press)

Charles T. Stewart Jr. (1958), 'The Size and Spacing of Cities', Geographical Review, 48 (April), reprinted in Readings in Urban Geography ed. Harold M. Mayer and Clyde F. Kohn (Chicago: Chicago University Press).

May 25th 1971

George J. Stigler (1951), 'The Division of Labor is limited by the Extent of the Market', Journal of Political Economy, 59 (June)

J. Tinbergen (1968), 'The Hierarchy Model of the Size Distribution of Centers', Papers and Proceedings of the Regional Science Association, 20

Raymond Vernon (1960), Metropolis 1985, (Cambridge, Mass.: Harvard University Press).

Karl Vind (1965), 'A Theorem on the Core of an Economy', Review of Economic Studies, 32 (January)

Georges Widmer (1953), "L'Inegalité dans la Grandeur des Villes et ses Correlations Economiques", Revue Economique.

A. G. Wilson (1970), Entropy in Urban and Regional Modelling (London: Pion)

Lowdon Wingo Jr. (1961), Transportation and Urban Land, (Washington D.C.: Resources for the Future, Inc.).

For Product Safety Concerns and Information please contact our EU representative GPSR@taylorandfrancis.com Taylor & Francis Verlag GmbH, Kaufingerstraße 24, 80331 München, Germany

T - #0159 - 160425 - C0 - 297/210/19 - PB - 9781138184862 - Gloss Lamination